D0402107

ALONE in the Universe?

Aliens, The X-Files & God

David Wilkinson

InterVarsity Press
Downers Grove, Illinois

InterVarsity Press
P.O. Box 1400, Downers Grove, IL 60515
World Wide Web: www.ivpress.com
E-mail: mail@ivpress.com

Published in the United States of America by InterVarsity Press, Downers Grove, Illinois, with permission from Monarch Publications, Great Britain.

InterVarsity Press® is the book-publishing division of InterVarsity Christian Fellowship/USA®, a student movement active on campus at hundreds of universities, colleges and schools of nursing in the United States of America, and a member movement of the International Fellowship of Evangelical Students. For information about local and regional activities, write Public Relations Dept., InterVarsity Christian Fellowship/USA, 6400 Schroeder Rd., P.O. Box 7895, Madison, WI 53707-7895.

Cover photograph: Tony Stone Images

ISBN 0-8308-1938-X

Printed in the United States of America

Library of Congress Cataloging-in-Publication Data

Wilkinson, David, Rev. Dr.
Alone in the universe? : the X-files, aliens, and God / David Wilkinson.
p. cm.
Includes bibliographical references and index.
ISBN 0-8308-1938-X (alk. paper)
1. Extraterrestrial life. 2. Interstellar travel. 3. Astronomy in the Bible. 4. Jesus Christ—Person and offices. I. Title.
QB54.W64 1998
261.5'13—dc21
97-47062
CIP

20 19 18 17 16 15 14 13 12 11 10 9 8 7 6 5 4 3 2 1
15 14 13 12 11 10 09 08 07 06 05 04 03 02 01 00 99 98

To Adam and Hannah,
That they may know they are not alone

Preface

Everyone who gives a public talk on astronomy or astrophysics expects one question. Although there may be a variety of questions, from "What is a black hole?" to "What makes the Big Bang go bang?" there seems to always be one hand that is inevitably raised to ask, "Are we alone in the universe?" It is a question that many lecturers dread, yet it is a question that is extremely exciting.

I do not believe that such a question is an attempt to show up the lecturer's ignorance, but is instead a fundamental question linking many different areas of science, philosophy and religion. If we are not alone, then what does that say about human beings and our place in the universe?

The question of whether we are alone in the universe is very popular in today's society. It is addressed from many sides, but often these sides work independently. The scientific community addresses the question with very little engagement with the claimed evidence of UFOs and alien phenomena. Those who major on the evidence for aliens visiting Earth address it with little regard to the scientific arguments of the evolution of life and space travel. The religious community, if it addresses the question at all, does so without real reference to the other two.

Yet here is a subject that can integrate these three perspectives. It seemed reasonable, therefore, as someone who is often asked this question, to attempt to look at it from all three angles in one book. There will be those who will claim that such a treatment is somewhat superficial, but it may have the advantage of sketching out the important issues that need to be addressed. It is often said that Giordano Bruno was burned at the stake in 1600 at the hands of the Inquisition for such a claim. In fact he was found guilty of multiple charges, of which an infinite universe and a plurality of other worlds were simply a small part. Nevertheless, an author needs to keep this in the back of his or her mind!

Many people have helped in the preparation of this book. It is a joy to thank Sir Arnold Wolfendale, who first raised the question for me and kept on raising it. Tony Collins has been a constant encouragement as to the value of this subject. Sir Robert Boyd, CBE, Sam Berry, Rob Gayton and Liz Gayton gave helpful comments on the manuscript. I am thankful for discussions with and help from Rob Frost, Neil Richardson, Kenneth Cracknell, Susan White, Brian Hoare, Brian Sibley, Wayne Clarke and Mark Kermode. Elm Hall Drive Methodist Church and Liverpool University Chaplaincy once again gave me the opportunity and support to write this book. My support group has been invaluable.

Finally, to my wife, Alison, and children, Adam and Hannah: thank you for your patience and love.

I am convinced that we are not alone in the universe. The following pages explain why.

1

Is the Truth Out There?

"Two-Year-Old Child Plays with Space Alien" is the kind of headline we expect to see in certain tabloid publications. Yet it is true of my household. A small toy version of ET is perched on top of the toy box, looking desperately as if he wants to go home. Elsewhere in the room Mr. Worf, a Klingon security officer with the USS *Enterprise,* stands ready to boldly go where no one has gone before along the mantel, and Buzz Lightyear, space ranger, battles to establish himself in the midst of the other toys.

Aliens have become part of the culture of our present-day world. Some, as the ones discussed above, are friendly, while others are not. From Ming the Merciless in *Flash Gordon* to the "Quick, let's hide behind the sofa!" Daleks of *Dr. Who* to the vile monsters who attack the Mighty Morphin Power Rangers, aliens both fascinate and terrify.

But do they really exist? Are these aliens simply the product of the human imagination that also created Pooh Bear and Tigger, or is there any possibility that such creatures actually do exist? We might say that if they do not exist, it would be a pretty boring universe. Can you imagine the crew on *Star Trek* unable to find any "new life and new civilizations"? But if these things do exist, then what are the consequences, not just for science but for our whole view of the universe?

If some of the recent scientific claims are to be believed, this fascination with aliens will be going strong for some years. And the implications for religious belief, some claim, are very serious indeed.

Fossilized Extraterrestrial Bugs

Life exists elsewhere in the universe! Or so it was claimed in the first week of August 1996, following the discovery of evidence from the planet Mars. Scientists claimed to have seen the fossilized byproducts of very primitive life within a meteorite believed to be from the surface of Mars. President Clinton hailed the discovery as "one of the most stunning insights into our world that science has ever discovered. . . . Even as it promises answers to some of our oldest questions, it poses others even more fundamental." We will return later to whether this meteorite did indeed contain a "stunning insight," but for the moment let us pursue the president's questions.

Some believe that these questions are theological and that the discovery of life elsewhere in the universe is another nail in the coffin of biblical Christianity. Paul Davies, the outstanding popularizer of science and professor of natural philosophy at Adelaide University, thinks so.

> Even the discovery of a single extraterrestrial microbe, if it could be shown to have evolved independently of life on Earth, would drastically alter our world view and change our society as profoundly as the Copernican and Darwinian revolutions. It could truly be described as the greatest scientific discovery of all time. . . . It is hard to see how the world's great religions could continue in anything like their present form should an alien message be received.[1]

Is this true? Would the confirmation of life on Mars be of any real significance? Are Clinton and Davies right to say that this discovery poses fundamental questions? Does God have anything to fear from life on Mars? Does life elsewhere in the universe contradict the Bible's view of the special nature of humanity? These questions are not new, but they are questions that are extremely popular in Western culture these days.

Long Ago in a Galaxy Far, Far Away . . .

The idea of other worlds and civilizations is at least 2500 years old and probably older. The Greek philosopher Epicurus (341-270 B.C.) was one of the first to suggest an infinite number of worlds and the existence of life in these other worlds. Even earlier, Plato (427-347 B.C.) argued that the maker of the universe "distributed souls equal in number to the stars, inserting each in each." He believed that if anything could be created it was created by the demiurge who built the universe. Plato's demiurge was not the sole Creator of all things, but simply an entity who worked with preexisting matter as an architect of the universe. It is interesting to note that such a view, as we will see, has its advocates today, although they express it in a different way.

The philosopher Democritus (450-? B.C.) provided the basis of this belief in the philosophy of atomism. According to this theory the world was made by the coming together of constantly moving "atoms." In the moving and coming together of atoms in an infinite number of ways, which was thought true for the whole of the universe, every possible outcome could be fulfilled. Life on Earth had arisen as just one example of many, as a result of atoms coming together. Since nothing was unique about this in the case of Earth, it was thought that this could happen in many different places throughout the universe. Such a philosophy in many ways still undergirds the belief in life elsewhere in the universe. It combines three assumptions about the universe:

1. The laws of nature are universal.

2. There is nothing special about Earth.

3. If something is possible nature tends to make it happen.

These assumptions often result in the conclusion that we are not alone in the universe.

If the possibility of other worlds has been philosophically accepted for a long time, many people have speculated about intelligent life on those worlds. The Pythagoreans, a century before Socrates, believed that the moon was inhabited by creatures vastly superior to those on Earth. This belief survived in scholarly debate until the eighteenth century. Plutarch (A.D. 46-120) first called the dark areas of the moon's surface "seas," and such names are still retained today. Apollo 11 landed on the Sea of Tranquility.

The great astronomer Johannes Kepler (1571-1630), observing the surface features of the planets and the moon, believed that the moon's inhabitants were stronger than we were because of the long, hot lunar days. He also argued from Galileo's observation that because Jupiter had moons, it too must be inhabited. He reasoned that as the moon had been made for our benefit by God, then the moons of Jupiter were made for the benefit of the inhabitants of Jupiter.

The same argument was used by Richard Bentley (1662-1742) in England and Christiaan Huygens (1629-1695) in Holland following the discovery of the vast number of stars in the universe. They reasoned that if there were stars that were unable to be seen from Earth, then they must have been created for the benefit of those who could see them. In addition, science historian Colin Russell has suggested that common to the many speculations about other worlds in the seventeenth century was an insistence on God's ability to create life anywhere he wished and that the universe existed not just for the sole benefit of human beings but to exhibit God's glory to all.[2]

Such belief in the existence of life elsewhere in the universe continued into the seventeenth and eighteenth centuries. When Benjamin Franklin, the inventor of the lighting conductor, heard in 1757 that the world might one day collide with Halley's Comet, he said, "We must not presume too much on our own importance. There are an infinite number of worlds under the divine government, and if this was annihilated, it would scarce be missed in the universe." Even as late as 1867 some astronomers still believed the sun had a cool interior populated by very bright people!

The Alien Revolution

How had such speculations become so respectable? Some four hundred years ago, prior to the Copernican revolution, human beings considered themselves to be the center of everything. The universe as described by Ptolemy (90-168) had Earth as its center and everything orbiting around in beautiful but increasingly complex circles. Men and women were the masters of it all. But in the past four hundred years, human beings have been dethroned by a succession of ideas.

Copernicus suggested that the Earth was not the center of the universe, and Galileo provided the observational support. Earth lost its special place and was relegated to orbiting the sun. Then Galileo and others showed that Earth is surrounded by many, many stars like our own sun. Our sun lost its special place and was shown not to be unique. This meant that there were possibly a great many planets around these other stars. Perhaps we were no longer unique as a planetary system. As astronomers understood more and more of our neighborhood in relation to the rest of the universe, it turned out that the star we orbit is pretty average; it is one star in one hundred billion stars about two-thirds of the way out from the center of the Milky Way galaxy, a galaxy fairly typical as galaxies go.

Along with these discoveries, science was uncovering laws that seemed to apply to every part of the universe. Newton's law of universal gravitation, which applied to apples dropping off trees, also seemed to work well with planets orbiting the Sun. The assumption further grew that the processes which led to life were universal as well.

In 1859 Charles Darwin's *On the Origin of Species* argued that we are here by the same process of evolution that led to plants and animals.

Natural laws were universal, and we were no longer special. Therefore it seemed quite reasonable to conclude that we were not unique. Life elsewhere in the universe simply followed the pattern of this dethronement.

However, at the end of the nineteenth century and into the twentieth, scientists began to become very skeptical about the possibility of other life. Such skepticism had always existed, even as far back as Aristotle (384-322 B.C.), who believed that our world was unique. Now science began to back up such a view. It became clear that the other planets and moons in our solar system were unable to support life. Evolution, which had given support to the idea of other life, was now seen as a very special process with a high degree of sensitivity to the conditions and environment. Life had developed here on Earth because of very special circumstances. Although planets around other stars might exist, it was thought unlikely that they were inhabited.

Today the cosmic pendulum is swinging back again. In the past decade there has been an explosion of interest in and belief in extraterrestrial life. No longer just the preserve of a small group of devotees, it has become a phenomenon that has invaded our culture in books, magazines and films.

I Want to Believe

Launched in September 1993 on the Fox Network, *The X-Files* has been a leading indicator of this renewed interest. It has been a phenomenal commercial success and is now a TV cult legend. It has made stars of Gillian Anderson and David Duchovny and spawned a feature-length movie, mass merchandising with T-shirts, books and comics, and even a Gillian Anderson Testosterone Internet Group. It has communicated to a mass market such subjects as alien abductions, visitations and elaborate government conspiracies.

The X-Files deals with more than just aliens, but aliens are an important part of a whole range of paranormal phenomena. FBI agents Dana Scully and Fox Mulder probe a series of cases, many of them taken from the folklore surrounding the existence and visitation of aliens. The "Erlenmeyer Mask" episode deals with the government's injecting alien DNA into humans, and in the "Deep Throat" episode the government wipes out the memory of an officer who has been involved in tests with alien spaceships. The them-versus-us theme of the government as Big Brother deceiving the public is central to the program and part of its appeal. Its multitude of alternative realities of aliens and supernatural experiences picks up a "truly global climate of star gazing fascination and world weary paranoia."[3] Chris Carter, the series creator, sees it in more fundamental terms, considering himself "a nonreligious person looking for a religious experience." A strategically placed poster in Mulder's office shows a flying saucer and the words "I want to believe."

Media Contact

Books such as Timothy Good's *Beyond Top Secret*[4] and Nick Pope's *Open Skies, Closed Minds*[5] have become bestsellers as they chronicle the evidence for an alien origin of unidentified flying objects (UFOs) and government coverups. Magazine racks are full of magazines with such titles as *UFO Magazine, UFO Reality, Alien Encounters* and *Sightings*. These magazines are very recent, many of them first published in 1996. Some are so certain of their market and content that they publish monthly. In comparison, the same retailers usually stock virtually no Christian magazines.

The Internet is full of information about aliens. You can browse through specific messages from certain aliens to extensive anatomical and psychological profiles of others. Once again, as might be expected in a medium that prides itself on open access to information, government conspiracy theories feature prominently.

Such interest is reflected in and continually fueled by science fiction. *Independence Day* became the biggest hit of 1996 and to date the fastest-earning movie ever. It took in $100 million in six and a half days, compared to the nine days required by *Jurassic Park.* Other Hollywood "alien" blockbusters followed such as *The Arrival* with Charlie Sheen; Tim Burton's comedy remake of *Mars Attacks!* with Jack Nicholson, Glenn Close and Pierce Brosnan; *Contact,* from the novel by Carl Sagan; the $97 million *Starship Troopers; Men in Black,* in which Tommy Lee Jones and Will Smith play cops on the trail of visiting aliens; Bruce Willis in *The Fifth Element; Alien Resurrection* (or *Alien 4!*), in which the ever-hungry alien meets a genetically resurrected Sigourney Weaver; a *Blade Runner* sequel; and Michael Crichton's *Sphere,* about an underwater mission to explore an alien spaceship.

The *Star Wars* trilogy, twenty years since Luke Skywalker took the controls of his X-wing T-65 spacecraft and polished off the Death Star, was released again in three special editions in 1997, while a new three-part trilogy goes into production.

Spock to the Kazons: Trekking Across a Crowded Universe

This continued interest in *Star Wars* grows alongside the phenomenal popularity of *Star Trek.* In *First Contact,* the eighth film of the series, released in the U.S. in November 1996, Captain Jean-Luc Picard and his crew encounter the Borg, a malevolent cybernetic life form intent on assimilating every other form of life into themselves.

First broadcast on September 8, 1966, the original *Star Trek* series lasted only three years before being withdrawn by NBC. However, in that time it built up a solid and loyal core of fans. One of the most interesting features of the first series was the character of the Vulcan science officer, Mr. Spock, played by Leonard Nimoy. NBC was very doubtful about having an alien on board, but he soon became the focus of public enthusiasm for and fascination with the series. Indeed, alongside the galactic soap-opera feel and imaginative technology, aliens became a central part of the show's appeal.

Star Trek introduced galaxies generously populated by alien life. From the aggressive Klingons to the cuddly Tribbles, the show painted a picture of a multitude of life forms and civilizations on every small and apparently innocent planet the *Enterprise* visited. In 1979 *Star Trek: The Motion Picture* took the series to Hollywood and told the story of the return of one of NASA's Voyager probes, given amazing powers by a distant world of aliens. This and the series of movies that followed grossed $1.3 billion worldwide. *Star Trek: The Next Generation* continued the story, now contending with the evil empires of the Romulans and Cardassians, as well as a new superrace called the Q.

The amazing thing is that all of these empires, plus the many individual civilizations and strange alien life forms, supposedly exist in just one part of the Milky Way galaxy! Another TV show, *Star Trek: Deep Space Nine,* has introduced another empire, the evil Dominion, through the mechanism of a worm hole linking different parts of the galaxy. And the latest *Star Trek* incarnation, *Voyager,* is stranded in yet another part of the galaxy with many more alien worlds to explore.

These fun and imaginative fantasies are at the opposite extreme of the scientific skepticism that says we are alone in the universe. They also raise the important question, If there is life out there, what will it be like and will it be able to communicate? In *Star Trek,* thanks to the "universal translator" all aliens speak in English! Rick Berman, executive producer of *Voyager,* comments, "We can come up with hundreds of different aliens, but the attractive thing about *Star Trek* is familiarity."

This belief in the existence of intelligent alien life in the galaxy that's not too different from us is reinforced by massive merchandising. Today you can buy everything from *Voyager* alarm clocks to Lieutenant Barbie of the USS *Enterprise* engineering section. A quarter of a million Klingon dictionaries have been sold, and degrees are now offered in Klingon. Indeed, there are heated "scholarly" debates such as whether in the Klingon translation of the Bible, "bread" should be rendered *Rokeg blood pie!*

It is clear that global consumption of science fiction appears to be at an all-time high. No doubt this is partly due to the power of merchandising coupled with the digital special-effects revolution where anything is now possible in movies. However, it also reflects the recurring question of whether there is other life in the universe, a question that modern science is taking more and more seriously.

A Question Beyond Belief?

Recently the popular science magazine *Focus* compiled its "one hundred weirdest mysteries known to science." Alongside questions such as "What are the ultimate building blocks of matter?" and "Where do all those odd socks go to?" the number-one mystery was "Are we alone in the universe?" Likewise commenting on plans to look for planets capable of sustaining life, NASA administrator Daniel Goldin said that public excitement about this field "is beyond belief."

This is an extremely important time for the scientific understanding of the existence of other life in the universe. It is only in the past few years that scientists have seen planets other than those in our own solar system. Better understanding of chemistry and biology are opening up the question of the origin of life. New projects to search the sky for signals from other civilizations are currently underway, and our space exploration programs take us to our nearest neighbor planets to search for signs of life.

It is fair to say that the claim of life on Mars has been well used by NASA for its own purposes. Since the Challenger disaster NASA has been struggling for direction and funding. In particular, missions to Mars, including a possible manned landing, have been in considerable doubt in recent years. But the claimed discovery of life on Mars, as well as the successful and relatively inexpensive Mars

Sojourner rover mission, has enabled NASA to trigger massive world interest and open the door to new funding.

Alongside this interest in the scientific mainstream there has also been a phenomenal growth of interest in UFOs and alien sightings. Some of this interest has led to some rather bizarre claims.

Strange Happenings

In 1995 police had to rescue three fortunetellers in Schtaklevo, Bulgaria. A crowd of fifteen hundred people had gathered on an airfield awaiting the landing of eight spaceships promising to bring $16 trillion from outer space. The money would wipe out the national debt, and there would be lots left over for those who had gathered. When the UFOs failed to arrive, the people grew angry at the fortunetellers, who claimed that the UFOs had been frightened away by jet fighters.

In many countries a small but significant "UFO community" has developed. This is a very loose grouping of people held together by books, magazines, lecture tours, conventions and contact via the Internet. Such communities consist of people who have either personally experienced UFOs and aliens or believe in them. Some of these reports are quite serious and have no obvious alternative explanation.

Some stories take on bizarre proportions and are not all accepted by the community itself. However, these claims are often the ones represented in the media, such as the idea that Hitler was in contact with aliens, that the Earth is in fact hollow and inhabited by a strange race of aliens, and that humanity evolved as part of an alien experiment in genetic engineering.

Such interest has produced a number of movements that mix UFOs, aliens and religion. The International Raelian Movement was founded by racing driver Claude Vorihon in 1973. He told of how he encountered an alien who appointed him as prophet to the world. The Movement is currently trying to establish an alien embassy with a landing pad for visitors from outer space.

In 1914 London taxi driver George King was washing some plates when he heard the voice of the Interplanetary Parliament, which appointed him as a prophet. King channeled messages from alien beings and apparently defended the world against a plot by the Evil Fish Fiends from Garouche to suck out the atmosphere of the Earth. Recently the organization he founded, The Aetherius Society, arranged for an alien craft to fly over the city of Los Angeles. But to avoid widespread panic the craft was made invisible and only a few believers saw it. This sort of stuff can easily be discounted, but for some it is very attractive. For example, Ray Davies of the rock band The Kinks has been involved with this society.

In the late 1940s a group of science-fiction writers met together to discuss what could be a successful modern religion in contrast to the traditional religions. One of that group was the well-known science-fiction author L. Ron Hubbard. Hubbard went on to found such a religion, the Church of Scientology, which is a rapidly growing religious movement. A successful method of psychotherapy, called "dianetics" by Hubbard, was combined with an "outer space" philosophy. Constructing his new religion, Hubbard used the mythology of science fiction and aliens. Each human body is inhabited by a spiritual, immortal entity called a Thetan. The Earth is a prison planet where Thetans are being dropped by flying saucers from wars in the Galactic Federation. The distinction here between science and science fiction is blurred, but it provides a powerful mix.

Often such a mixture of aliens and religion is given credibility through some science and the images of science fiction. The Church of Scientology has been in the news recently due to questions about whether it is a charity or a cult and because of a campaign in Germany to boycott the movie *Mission: Impossible* because of actor Tom Cruise's involvement in the church. Viewed by many as a cult, the Church of Scientology counts a number of leading Hollywood stars among its members, most notably Tom Cruise and John Travolta.

Heaven's Gate

Sometimes this powerful mix leads to serious consequences. Comet Hale-Bopp, named after the two amateur astronomers who discovered it on July 23, 1995, was probably the most viewed comet of recent history. Unlike Halley's Comet, the disappointing "smudge in a telescope," Hale-Bopp sat in all its glory for over a month high in the evening skies over the Northern Hemisphere. Despite its brightness, it came no closer than 120 million miles of Earth.

Such a distance from Earth was very safe for us. Many theories abound as to whether comets bring death or life. Some argue that a comet impact was responsible for the catastrophe 65 million years ago that led to the extinction of the dinosaurs. Others see comets bringing to planets like our own the carbon-based molecules necessary for life. As we will see later, this theory exists not just to explain the difficulties with evolution, but because the gases released by comets contain organic compounds.

However, few would have expected that Hale-Bopp would claim thirty-nine human lives. Their bodies were found in a $1.6 million mansion near San Diego on March 26, 1997, in what seemed to be a religious mass suicide. Laid out on their backs on bunk beds and mattresses, dressed in black, faces hidden by purple shrouds, they had left identification papers and goodby messages on videotapes and the Internet. They had killed themselves through a combination of drugs, alcohol and plastic bags over their heads.

They were all members of a cult led by Marshall H. Applewhite, a former music professor. Fired from his job for a homosexual affair, his life changed in 1971 when he founded the movement with his nurse Bonnie Lu Nettles. They became the "two," convinced that they were the two witnesses prophesied in Revelation to prepare the way for the "Kingdom of Heaven," a level above human existence.

What made this cult suicide different from the nine hundred followers of Jim Jones who drank cyanide-laced grape punch in the jungles of Jonestown, Guyana, or from David Koresh and seventy of his Branch Davidians in Waco, Texas, was the link to aliens.

Marshall Applewhite's followers took Comet Hale-Bopp as a sign to commit mass suicide, for they believed that there was an alien spaceship following behind the comet and using it a shield. They thought they were beings from another planet simply inhabiting the "containers" of their human bodies. Through committing suicide their immortal souls would be released and taken by the spaceship to the kingdom of heaven. They saw Hale-Bopp as "Heaven's Gate."

The Heaven's Gate cult was not alone in this kind of belief. Since 1994 seventy members of the Order of the Solar Temple have taken their lives in Europe and Canada. They believe that ritual suicide leads to rebirth on a planet called Sirius.

The Heaven's Gate cult displayed once again a powerful mixture of religion, science fiction and the belief in aliens. Their website attracted Internet surfers looking for anything from alien abductions to the Second Coming. They mixed end-of-the-world eschatology with a space-alien obsession, ridiculing Christianity but using biblical references along with their fascination with the aliens and terminology of *The X-Files, ET, Star Wars* and *Star Trek.*

Fascinating . . .

In order to examine the question of whether or not life exists elsewhere in the universe, we will discuss each aspect of the question in detail. In the next four chapters we will look at the scientific questions: What is the scientific evidence for or against life elsewhere in the universe? How common is life? How does it appear? Does evolution mean that all life converges to humanity, or are there many different forms? What is consciousness?

We will then examine the evidence for and against the claim that either in past history or in the present we have had direct contact with aliens. Following this we will see what problems for Christianity, if any, are caused by the existence of aliens. Finally, we will ask what elements are influential in supporting our interest in aliens.

The question of whether we are alone in the universe is profound. It unites the imagination of science fiction with the hard science of planet formation and biological evolution. For some it provides the soil for the root of a new religion, while for others it asks difficult questions of traditional Christianity.

2

Close Encounters of the Scientific Kind

I do not wonder whether we'll detect another civilization. I wonder when. The silence we have heard so far is not in any way significant. We still have not looked long enough or hard enough."[1] So wrote Frank Drake, a radio astronomer and a strong supporter of the existence of extraterrestrial life. In contrast, Marshall T. Savage is convinced that the odds against the appearance of life are just too great for the event to have occurred more than once. He comments, "The skies are thunderous in their silence; the Moon eloquent in its blankness; the aliens are conclusive by their absence. They've never been here. They're never coming here because they don't exist."[2]

Such a disagreement is characteristic of the divergence of the scientific community on the question of extraterrestrial life. Those who strongly argue for it tend to be astronomers and physicists.[3] Those who argue against it tend to be the leading experts in evolutionary biology. They suggest that because the emergence of intelligence is very unlikely, humanity is probably unique.[4]

The hub of the debate involves the following key questions:

- Is there any evidence for life on the planets in our solar system?

☐ Do planets capable of supporting life exist elsewhere in the universe?

☐ How does life begin, and how does it evolve?

The first two questions we will address in this chapter. The third we will leave for the next.

Mars Attacks!

Allan Hills 84001 is not a new soap opera but a very important piece of rock. Weighing 1.9 kilograms and found in Antarctica in 1984, it may give us a clue to the existence of life elsewhere in the universe. Within it are pockets of glass that carry the same gases that make up the atmosphere of the planet Mars. They are sufficiently different from our own atmosphere to suggest that the rock itself was once part of the Martian surface.

It is one of eleven such meteorites that are believed to have come from Mars as a result of huge explosions on the planet's surface. Material from Mars is in fact quite abundant. Some five hundred tons fall on the Earth each year. In 1911 a piece of Mars known as the Nakhla meteorite fell to Earth and killed a dog in Egypt!

Some fifteen million years ago the impact of a comet or asteroid on the surface of Mars threw up many rocks, some of which escaped the gravitational pull of the planet and went wandering through the solar system. The atmospheric gases of the planet were trapped on the surface of the rock when the asteroid hit. Some thirteen thousand years ago one of these rocks (ALH84001 for short!) entered Earth's atmosphere and landed in Antarctica.

In August 1996 a team of NASA scientists led by David McKay published their claim to have found evidence inside the meteorite of long-dead bacteria. They identified fine-grained magnetite and iron sulfide particles that are similar to those produced by bacteria on Earth, as well as tiny spheres of carbonate materials that are further evidence of biological byproducts. The media were presented with pictures showing wormlike structures no more than a hundredth of the diameter of a human hair, with the claim that these were fossils of Martian bacteria. The *Daily Mail* called it "virtually nothing but a vague orange-colored smudge."

This "smudge" led to worldwide headlines screaming "We are not alone!" and even comments by President Bill Clinton. But we need to be clear about the facts. Despite the headlines in the media, little green men and women have not been found on Mars. However, there may be fossilized leftovers that could have been produced by primitive life.

Yet even this is controversial. As the science team itself admitted, the evidence so far is not compelling proof. In the journal *Science* the team admitted that every feature of ALH84001 can be explained by itself, without the idea of life on Mars.[5] For example, some have suggested that the wormlike structures could have been made by dried clay. Nevertheless, the team argues that all of the evidence taken together points to their conclusions.

Some scientists have agreed with the evidence, while others have flatly rejected it. Monica Grady, curator of meteorites at the Natural History Museum in London who has studied ALH84001, said, "I am completely unconvinced there is any evidence on this meteorite to support the idea that life once existed on Mars." Even more important, although this did not grab the headlines, is evidence that these structures were formed at extremely high temperatures. This would point to another formation mechanism besides bacteria.

In addition, even if these are fossil bacteria, can we be fully sure that they evolved on the surface of Mars? It is not impossible that the bacteria could have gotten into the rock during its time on Earth, although the fact that they are deep inside the rock makes this unlikely. A more realistic possibility is that the rock was on its second leg of a round-trip ticket to Mars. The rock could have initially been ejected from Earth by the same mechanism that ejected it from the surface of Mars. Microorganisms can survive quite lengthy journeys in space, provided that they are concealed deep in rocks. We may be seeing simply a primitive organism that evolved on the surface of the Earth and then went on its own space journey.

A Martian Chronicle

Of course, claims of life on Mars are not new. In 1877 the Italian astronomer Schiaparelli reported the existence of dark lines on the surface of Mars. He used the word *canali* for these lines, which was misunderstood in English as "canals." As a result, one hundred years ago Percival Lowell built his own observatory in Arizona to look for life on Mars. He observed the surface of Mars and saw patterns that changed. He identified such changes with life. The belief grew that "canals" had been built to bring water from the polar icecaps to the vegetation of the equatorial regions. Unfortunately, such beliefs were wrong. The lines were not made by intelligent life.

Some twenty years ago NASA sent the Viking spacecraft to Mars. Not one single organic molecule was detected, even though the instruments were sensitive enough to detect one part in a billion.

It is interesting that such claims of fossil life were made thirty years ago about a meteorite called Orgueil, after its place of discovery in France. Headlines once again told the news around the world. The "fossils," however, turned out to be made of furnace ash, while the supposed organic material consisted of grains of ragweed pollen. Another meteorite, the Murchison meteorite, which fell in Australia in 1969, was again the source of a claim by Sir Fred Hoyle that it contained a fossilized single-celled organism. These "discoveries" simply remind us that the claims of the NASA team in 1996 need to be looked at with great caution.

Mission to Mars

Further scientific evidence should answer some of these questions. A new group of experiments on meteorite ALH84001 will look for cell walls and amino acids. This will be followed by a series of probes that will visit Mars in years to come. On July 4, 1997, NASA's Pathfinder Project placed a lander and a six-wheeled robotic rover on the planet's surface. The photographs of the Martian surface were stunning, and the technical feasibility of reasonably low-cost missions to Mars was proved. Pathfinder is the beginning of a ten-year series of space probes to be launched to look for fossil evidence. These missions will be able to drop landers and weather stations onto the Martian surface, which will analyze the soil chemicals and the layer of permafrost believed to lie beneath the ground. But if life does exist, it is most likely to be much further down.

Even if life did begin on Mars at the same time as it did on Earth, some 3.8 billion years ago, as Mars started to freeze living organisms would have retreated into rocks or into the planet's interior, seeking warmth from volcanic hot springs. The surface conditions are just not able to sustain life. Mars is too dry, any water being frozen in permafrost, and intense ultraviolet radiation sterilizes the surface. The Viking missions' negative results for life on the Martian topsoil are consistent with this.

However, three billion years ago there was water on Mars, which left its mark in channels and valleys on the surface. This could have enabled some organisms to form. The discovery of primitive forms of life here on Earth that live in volcanic hot springs and around ocean thermal vents has been important in suggesting that life can evolve and be sustained in rather harsh conditions. Such areas have a wide range of chemicals and enough energy to ensure rapid reactions. Heavy metals are found here, whose large atoms would be expected to lead to the rapid formation of new complex molecules, and there are clays whose unique shapes might produce templates for the formation of these complex molecules, which could lead to life. The creatures that live in such places are called thermophiles, and some of these are known to be 3.5 billion years old. These primitive creatures occupy many of the lowest branches on the tree of life. They and their relatives are among the direct ancestors of today's animals and plants.

It seems possible that such conditions existed on Mars, so life could have developed in a similar way. As time went on, the organisms may have retreated deep into rocks or deep below the surface. The real difficulty is that to find them the Martian probes would have to drill anywhere from three hundred feet to almost a mile under the surface. So far this lies beyond the ability of even the planned Mars probes. The first landing and retrieval mission is planned by NASA for 2005, but even then results may not be conclusive.

It's a Long Way from Bacteria to Little Green People

What then would be the importance of finding primitive life on another planet that had evolved independently of life on Earth? In the case of Mars, if confirmed it would suggest that the process that forms life is widespread throughout the universe. To find evidence for life on one of our closest neighbors that has a very harsh environment means that life can develop in far more diverse circumstances than we had imagined. It is rather like coming back from a shopping trip having being told by the seller that the article you have purchased is rare and difficult to get—only to find that all of your neighbors have the article as well!

However, there is another important lesson from Mars. If primitive life is confirmed, it also shows that life does not necessarily develop to intelligent life. We will need to return to this question throughout the book. Science fiction, to be interesting, always needs to deal with intelligent, self-conscious life out there in the universe. If life did begin on Mars, it did not develop into little green women and men. In fact it did not develop beyond even the most primitive form. This is a reminder that circumstances are important in the development of life, but even more important in the development of life forms that we humans might ever have the possibility of relating to.

Men Are from Mars, Women Are from Venus

If it is possible that primitive life may have formed on Mars early in its history, what about the other planets in our solar system? Venus is our next-door neighbor, only twenty million miles away. Ancient cultures associated Venus with beauty, and there are even stories that the founder of Scientology, L. Ron Hubbard, claimed to have visited there. Venus features in a great deal of science fiction. It is difficult to see what the surface is like, which is shrouded in clouds, and this encouraged the belief that it might have inhabitants.

The trouble is, however, that such inhabitants would feel like they were living in hell. A series of space probes culminating in the Magellan spacecraft, which arrived in orbit in August 1990, have given us quite a detailed picture of conditions on the surface. Far from being a beautiful place, Venus is so hot that lead would melt, the atmospheric pressure is ninety times that of the Earth, and it has white clouds of sulfuric acid which produce interesting rain, to say the least. These conditions make the development of even primitive life extremely unlikely.

Looking elsewhere in the solar system does not contradict such a pessimistic view. Take Jupiter, for example. At 475 million miles from the sun, it is more massive than all the other planets put together. Some thirteen thousand Earths could fit inside it. It drags around a multitude of moons, two of which are larger than our moon. The nature of Jupiter has been revealed by ground-based telescopes and a series of probes, most recently the six-year mission of the Galileo probe.

Jupiter, like Saturn, Uranus and Neptune, is a gas giant. No one could walk on its surface because its atmosphere extends downward for tens of thousands of miles. The Galileo probe got down only one hundred miles before being crushed by the pressure of gases. Deep within, there could be a solid surface of hydrogen, but no one knows for sure.

Some suggestions of life on Jupiter have been made, but it would have to be quite an exotic form of life, perhaps able to float in the atmosphere. The moons would be a more likely place to find life, although the largest moon, Io, has a dozen active volcanoes and is covered with thick sulphurous clouds. Europa, another moon, has no atmosphere, is icy and very smooth. There have been some recent claims that Europa's icy surface could be floating on slush or even water. That water does exist in liquid form is a prerequisite in the eyes of many scientists who study the formation of life. However, the temperature of this moon that is so far from the sun makes it unlikely that any form of life might have developed.

The Goldilocks Principle

But Jupiter does highlight the odd circumstances that allow us to exist. In 1994 comet Shoemaker Levy 9 collided with Jupiter in such a spectacular fashion that scientists and the media were delighted. The important thing to note is that if Jupiter had not been there, the comet could have collided with one of the inner planets, perhaps Earth.

The Oort cloud is a cloud of perhaps millions and millions of comets that orbit the edge of the solar system. Some of these comets are occasionally knocked out of this cloud and into orbits that take them into the inner solar system. This is somewhat serious for us, especially if the Earth's orbit intersects the orbit of the comet!

Such a comet impact was a possible cause of the extinction of the dinosaurs some sixty-five million years ago. A comet impact today could have a similar effect, causing perhaps a billion people to perish. In fact some scientists take this possibility so seriously that they have been pushing for the world to set up an extensive program to spot incoming comets or asteroids and then have the technology in space to destroy them.

Without Jupiter the situation would be far worse. The huge planet considerably reduces the number of potentially lethal comets. George Weterill of the Carnegie Institute of Washington states,

> Without a Jupiter sized world in our planetary system, collisions with large comets and other dangerous objects like massive asteroids might occur with terrible frequency, not once every 50 million years as they do at present, but at least once every 100,000 years. This would make it extremely difficult for a civilization to evolve, and the simple answer is that there might not be one.[6]

This is yet another reminder of the very special circumstances needed for intelligent life to evolve. This has been called the Goldilocks Principle, because the many conditions need to be "just right." Stars have a region of space around them where conditions favor the emergence of life. If the temperature is too high, then living tissue would be destroyed, and if too low, then normal chemical reactions would not proceed. And as Jupiter shows us, the rest of the planetary system needs to be configured in a particular way in order for life to exist on one of the planets in the system.

To Infinity and Beyond!

There is a very simplistic argument for the existence of life elsewhere in the universe. There are 100 billion stars in each of 100 billion galaxies. This means that the number of stars is so huge that there must somewhere be a star like ours with a planetary system like ours on which life has developed. In fact some suggest that there are one thousand stars like our sun within one hundred light years of Earth, and that these would be ripe for life.

That sort of argument sounds convincing, but we need to be careful. First of all, not all stars, even if they had planetary systems, could support life. If a star is much larger than our own sun, it has a much shorter lifetime. These stars burn their hydrogen into helium and pump out energy far quicker than our sun. Apart from the higher temperature and level of radiation, such a star might have used up its available hydrogen fuel long before life had developed. If a star is much smaller than our sun, then any planetary system around it would be much cooler than our own, and this causes problems too.

Second, between 50 and 75 percent of stars are born with one or more companion stars. These binary or multiple systems of stars mean that it is difficult to see how planets would easily form. In the formation process a companion star depletes the dust and rocks that would normally form planets. And there is an even more serious problem for life in such a system. If planets had indeed formed, one of the stars might use up its fuel and then undergo a supernova explosion, becoming a neutron star or a black hole. The supernova explosion would send shock waves and intense radiation through the planetary system. And if that were not sufficient, then the radiation from the remnant neutron star or black hole would finish off the job. On the basis of this, solar systems may be the exception, not the rule, in our galaxy.

Third, we need to ask just how many planets are out there. Putting aside multiple systems and stars of the wrong size, how do planets form around a star like our sun, and is there any evidence of the process occurring elsewhere?

With or Without Planets?

About 4.5 billion years ago a vast cloud of gas, trillions of miles in diameter, began to collapse under its own weight. This cloud was part of a giant complex of clouds composed mainly of molecular hydrogen. Also in the cloud were relatively small quantities of many other elements such as carbon, oxygen and iron. These elements had been produced in the death throes of a previous generation of stars and had been spewed out into space.

The gas at the center, as it was compressed, heated up until, at a temperature of 30 million Kelvin,[7] hydrogen was fused into helium and a star was born. The star had a violent outburst that blasted the lighter gases on the edge of the cloud out into the rest of space. The system was now like a rotating fried egg. The star at the middle was surrounded by a rotating disc of different chemical elements. It is these "off cuts" that began to form planets. Complex mechanisms meant that the heavier elements such as iron coalesced into rocky inner planets, while the lighter gases formed giant outer planets.

The third planet from the star seemed to have a mixture of special circumstances. It held a stable orbit at a distance from the star where its surface temperature meant that water existed in liquid form. It was of sufficient size that it was able to retain an atmosphere that a smaller body like its own moon was unable to do, and it was able to sustain a reasonable level of geological and meteorological activity that made the existence of life possible.

So goes the history of Earth, or at least the story that the majority of scientists hold. Yet some parts of the story are still unclear, for example, does the collapse of the cloud just happen, or does it need a trigger such as the shock wave from a nearby exploding supernova?

The formation of planets should be commonplace if the history of Earth as told above is typical. Essentially, whenever a star is formed, planets form with it. There are a lot of stars in the universe, so there must be a lot of planets.

However, there is not universal agreement on this. One or two dissenting voices hold that planets do not form automatically when stars form. For some years now there has been an alternate theory, that planets are formed in the rare event of two stars passing close by each other. All stars are moving relative to one another. This leads very occasionally to two stars coming close together. When this happens, as the theory goes, some of the material of one of the stars is ripped out, and it is this material that eventually forms planets. As this is a rare occurrence, planets are a rare breed in the universe.

Although most physicists favor the first model rather than the second, the real test is whether we can observe any other planets outside our solar system. The difficulty with any scientific theory is building it on only one piece of evidence. If planets seemed to be abundant outside our solar system, then this would provide conclusive evidence for the first theory rather than for the second.

The difficulty is in actually seeing other planets outside our own solar system. Up until five years ago we knew only our own planets. The main difficulty was that stars emit a billion times more light than even the largest planets such as Jupiter. It is like picking out a light bulb beside a searchlight. But we actually began to see the light bulbs.

Finding the Silent Planets

One of the most exciting discoveries since 1994 has been the evidence of planets outside our solar system, a startling breakthrough. Until this time we had no evidence at all that the Sun was not alone as a star that had a number of planets.

Astronomers have been attempting to find other planets for many years. Acknowledging that the intensity of light from a parent star would make it difficult to see planets directly, astronomers developed indirect techniques. They attempted to look for the influence of planets on their parent stars. As a planet orbits around a star, the star "wobbles" in its position due to the gravitational pull of the planet. This wobble has an effect on the light the star emits. These changes have been difficult to measure, and any real results have taken some time to appear.

However, in 1994 Alexander Wolszczan of Pennsylvania State University provided evidence of three planets orbiting pulsar PSR 1257+12. The pulsar is only about six miles across but contains more matter than our sun. It spins rapidly and emits a beam of radio waves rather like a lighthouse. The three planets cannot be seen directly, but they change the period of the radio pulses as they orbit around the pulsar. The pulsar itself formed as the leftover remnant of a star that underwent a massive supernova explosion. This explosion would have destroyed any planetary system the star had at the time, so the planets that are now seen are thought to have been formed from the debris of a companion star also disrupted by the pulsar.

This planetary system is strange, and there is almost a zero possibility of life here. Indeed, any life on the planets, having survived a catastrophic explosion, would find itself now living beside a gigantic x-ray machine! What is important about this system is that it was the first confirmation of planets of any type outside our solar system. And through the work of two other astronomers, this system provided some even more intriguing results.

In 1995 Tsevi Mazeh and Itzahak Goldman of Tel Aviv University suggested that there might be a universal mechanism to planet formation. They pointed out that the three planets orbiting pulsar PSR 1257+12 have properties that almost exactly match those of Mercury, Venus and Earth in our inner solar system. It may be that there is a law of nature that ensures planets always have certain sizes and certain orbits.

It is possible to calculate the masses and the distances of the planets around the pulsar. The two outer planets are roughly the same mass, but the inner planet is much smaller. This is not dissimilar to the small Mercury and similar Venus and Earth, although it could simply be coincidence.

More surprising is that the ratio of the distances of the planets from their central pulsar seems to roughly follow Bode's Law, a numerical law concerning the ratios of distances of the planets in the solar system. If you take the sequence of numbers 0, 3, 6, 12, 24, etc. (each number after 3 is twice the previous number in the sequence), add 4 to each number and divide by 10, you get 0.4, 0.7, 1, 1.6, 2.8, etc.

These are the distances of the planets from the Sun in terms of the distances to the third planet, Earth. It has never been understood where these ratios come from, and many astronomers have dismissed this law as mere coincidence. Now the intriguing thing about the planets of PSR 1257+12 is that the distances of the planets from the pulsar in terms of the distances to the third planet are 0.4, 0.77, 1.

This is quite a coincidence, with profound implications. It could mean that Bode's Law, rather than being just a neat game with numbers and planets in our solar system, could be expressing a deeper universal law of planetary formation. And if this law applies to both our solar system and a pulsar, then solar systems like our own may be the rule rather than the exception throughout the galaxy.

Nevertheless, astronomers were still looking for planets around a "normal" star. The wobble of a star produced by the presence of planets can produce subtle changes in color, from blue to red, in the light the star emits. This change in the color of the light is called Doppler shift and was the basis for two more exciting discoveries.

Planets Capable of Harboring Life, or at Least Hot Tea?

In October 1995 Michel Mayor and Didier Queloz of Geneva Observatory detected a planet circling the star 51 Pegasi, which is forty-five light years away in the constellation of Pegasus. They estimated that it was about half the size of Jupiter but closer to its star than Mercury is to the sun. The planet takes four days to orbit the star and could have a temperature of around 1000 Kelvin. With this extreme heat it is likely that there would be no atmosphere and the planet would simply be a very large rock! This was the first planet found around a normal star, and it added to the sense that planets were widespread in the universe. However, such a large planet so close to its star does not seem to fit with Bode's Law and the theory of Mazeh and Goldman. It is not surprising that planet formation may be a little more complicated.

Then in December 1995 Geoffrey Marcy and Paul Butler of San Francisco State University discovered what they believed to be a planet around the star 70 Virginis, which is around fifty light years away. Early in 1996 headlines read, "American Astronomers Discover a Planet Capable of Harboring Life." Once again it seemed that life in the universe could be abundant.

In fact this and another planet discovered by Marcy and Butler are not the most hospitable of planets. The one orbiting the star 47 Ursae Majoris, some thirty-four light years away in the constellation of the Big Dipper, is at least twice the size of Jupiter and probably has the same chemical composition of gas. This is a rather unhealthy mixture of hydrogen sulfide, ammonia and methane. The other, which orbits the star 70 Virginis, is six times the mass of Jupiter. It too is a gas giant with all the problems of sustaining life. It also has a rather egg-shaped orbit, which has led some to believe that it is a small star rather than a planet.

So why the headlines? Why was it claimed that planets with possible life had been found? The reason is that the astronomers were able to calculate the temperature of both planets, and both are temperate enough for water to exist in liquid form. They are about the temperature of hot tea, which seems to be essential for life to form. Liquid water is an ideal medium in which carbon-based organic chemicals can dissolve and react with one another.

Bursting with Life?

However, we need to stress that even these discoveries do not confirm other life in the universe. The media hype that built the possibility of life from the calculation of a planet's temperature is rather like concluding that soccer is the national sport of a country after being told it has a lot of green fields. Many more things need to be examined before such a conclusion is drawn. Even the most optimistic believers consider that life on the newly discovered planets would have to be very exotic, existing in the upper atmospheres where the water might be in liquid form.

Some are prepared to push the argument further. They note that our own solar system has smaller planets as well as the large gas giants such as Jupiter. If our solar system is a typical example of the formation of planets, then we might expect that the large planets of 47 Ursae Majoris, 70 Virginis and 51 Pegasi should have other smaller planets with them, and if not planets, then perhaps moons that could sustain life. Michael D. Lemonick writes,

Perhaps most important of all, the discovery of planets around relatively nearby sun-like stars implies that our galaxy, the Milky Way, 100 billion stars strong, must be bursting with other worlds and that there is life out there somewhere.[8]

The argument has gone from some initial scanty observations to a positively Star Trek proportion in terms of life elsewhere in the galaxy. It is clear that far more observations are needed both to see other planets and more importantly to study their atmosphere and composition.

Much work can be done in this area if funding is provided. Scientists await results from a European satellite, the Infrared Space Observatory (ISO), that can detect heat from faint planets. In 1997 a new infrared camera was installed on the Space Telescope that could take a picture of one of the newly discovered planets. A new NASA initiative, the Origins program, has the goal of understanding the origins of the planets. In 2002 advanced cameras will be flown to the Space Telescope. Using techniques that suppress the glare of a star, they would reveal its planets, if any.

New space telescopes put on the moon or in deep space could search the planets' atmospheres for the presence of chemicals such as oxygen and carbon dioxide, which are needed and produced by living beings. In 2010 NASA hopes to launch Planet Finder, a telescope with five mirrors, each three to six feet in diameter, spread out over three hundred feet and placed in an orbit by Jupiter. The information collected by these mirrors would be combined electronically to simulate a much larger telescope. This should cost several hundred million dollars but would give us the information about the planets we very much need before conclusions can be drawn.

Planets, of course, are only the starting point. Although the possibility of vast numbers of planets throughout the universe increases the chances of finding intelligent life, we need to go beyond that and ask some fundamental questions about how life itself develops.

3

It's Life, Jim, but Not As We Know It

In 1953 Stanley Miller and Harold Urey at the University of Chicago performed one of the most famous experiments of all time. They attempted to produce in the lab the conditions on the surface of the Earth some four billion years ago. They passed an electric discharge through a mixture of water, hydrogen, methane and ammonia for several days. The liquid turned reddish brown and was found to contain several amino acids, the building blocks of all living organisms on Earth. This was an amazing result, suggesting that life could arise spontaneously with the right chemicals and the right conditions.

Although that experiment is now questioned as to whether it was a fair representation of the kind of conditions existing on primitive Earth, it sets the stage for the widely accepted standard picture of the development of life. It raises some very important questions.

In terms of alien life, two contradictory conclusions have been built. Those who believe in life elsewhere in the universe stress how "easily" life can develop. They claim that because the Miller-Urey experiment shows that the processes of life can be easily reproduced, such processes have been going on throughout the universe. On the other hand, those who believe that we are alone stress the importance of having the "right conditions" and that what was formed in the experiment was a complete fluke.

Who is right?

From Ashes to Amoeba

It is a big step from Miller and Urey's amino acids to human beings. Since the pioneering work of Sir Fred Hoyle and others in the 1960s, there is good evidence that atoms such as carbon, which are needed for life, were produced in supernova explosions earlier in the history of the galaxy. It is often said that human beings are made from the ashes of dead stars, or as Carl Sagan put in the television series *Cosmos,* "We're all made of star stuff." Such statements may be factually correct, but they do not mean that human beings are just star ashes or that the emergence of human beings from such ashes is a well-understood process.

First of all, although Miller and Urey may have made amino acids in their laboratory, there remain questions about how such organic building blocks could be produced in such large numbers and with sufficient concentration on the surface of the Earth. Geologists find microfossils of single-celled organisms in rocks that are about 3.5 billion years old. This means that amino acids must combine to produce life reasonably quickly in terms of the age of the Earth.

The critical step from abiological molecules such as amino acids to something like a one-celled organism is a giant one that is not well understood. Some might say that the amino acids just get together by chance in the same way as they themselves were produced.

This may sound like a reasonable argument until one realizes what is involved. Each living cell contains large molecules such as proteins and the nucleic acids, DNA and RNA. Each protein consists of different sorts of amino acids put together in a very specific order. There are about twenty different amino acids, and a typical protein will contain around one hundred of these, which have to be arranged in an exact sequence for the protein to work.

DNA and RNA are even more complex. These nucleic acids are made up of long chains with upward of tens of thousands of four different sorts of nucleotides. The sequence of nucleotides in the chains is the genetic code, the basic information that a cell needs to function and reproduce. The arrangement of these nucleotides on the DNA can be copied onto the RNA, which acts as a messenger, and then used by the protein-making machinery of the cell to produce the exact sequence of amino acids in each protein. So in the context of the origins of life, there follows the question of how the DNA and genetic code are produced. How are such long and complex chains created?

There is a further problem. The proteins that are made under the instructions of the DNA code are required for all of the functions of the cell. This includes the synthesis of the nucleotide building blocks needed for the production of the DNA itself. It is a classic "chicken and egg" problem. Proteins cannot be synthesized without DNA or RNA, and DNA cannot be synthesized without the proteins acting as catalysts in the building of the nucleotide chains of DNA. As Sir Karl Popper has written, "Thus the code cannot be translated except by using certain products of its translation."[1]

In light of this we are forced to the conclusion that although the building of basic amino acids may have happened spontaneously, the origin of life is very improbable in terms of its spontaneous appearance. Far greater complexity is needed, and that complexity must be of a certain kind in terms of specific chemical forms and reactions. Sir Fred Hoyle once commented that the formation of life by accidental molecular shuffling was like a whirlwind passing through an aircraft factory and assembling a Boeing 747 from the scattered components!

Indeed, many have estimated that the odds for random interactions to produce living cells in this way would be between 1 in 10^{30} and 1 in $10^{40,000}$. Such figures have an immediate implication for estimating whether there is life elsewhere in the universe.

If we assume that planets form when stars themselves form, which probably is the case (see chapter two), then there are only between 10^{19} and 10^{24} planets in the universe. Comparing these numbers with the chances of life forming spontaneously, it is clear that life should only exist on one planet within the observable universe, Earth. This is the traditional view, that life is a chance combination in unique circumstances.

They Came from Outer Space?

In response to such numbers and the mystery of the origin of life, a few scientists have suggested a very different theory, which is often known as "panspermia." It is that pre-Darwinian molecular evolution took place in space and was then transported to and seeded fertile planets like Earth.

Since the middle of the nineteenth century it has been known that certain rare meteorites contained organic chemicals. We now know that gas clouds in between some stars are composed of a rich variety of organic molecules. A gas cloud was recently seen around the newly formed star G34.3, some ten thousand light years away. From analyzing its spectral lines scientists discovered that this cloud, which is bigger than our solar system, has enough alcohol in it to make 4×10^{30} pints of beer, or 300,000 pints for every person alive every day for the next billion years!

At the forefront of this theory are Sir Fred Hoyle and Chandra Wickramasinghe.[2] They argue that life might have developed in its most primitive forms in space, and it was then transported by comets onto the surface of planets where it then began to evolve. Alternatively, there would be times when the solar system passed through a gas cloud and primitive organisms from the cloud entered the atmosphere of the planets. The implication of this is that life is widespread throughout the galaxy in molecular gas clouds, just waiting to land on a suitable planet.

Another scientist following this theory, although perhaps going further, is molecular biologist Francis Crick, who won the Nobel Prize for his work on the structure of DNA. His view is that life did not originate on the surface of Earth but was sent here by some intelligence. He suggests that microorganisms were sent here in an unmanned space rocket by a higher civilization billions of years ago.[3]

Although such a theory gets around the problems of life originating spontaneously, it creates other problems: How did the living cells originate in outer space, with the extremes of temperature and radiation? If they were sent by another intelligence, then where did that intelligence come from? And how did that life first develop?

Some have argued that the recent claim of primitive life on Mars would give some support to this theory. Life could have been transported to Mars in exactly the same way it came to Earth, but the conditions on Mars were just not right for it develop the way it did here.

Or It Came from RNA?

However, life on Mars, if it is shown to have arisen, could be interpreted differently. Maybe there is some mechanism, not yet understood, that formed the first living cells. Such a mechanism must be based on the idea of the DNA molecule evolving. Some suggest that a smaller RNA molecule of about eighty nucleotides appeared first, and this is how life developed. RNA has certain forms that can act as a catalyst instead of the protein. Laboratory experiments can get very short RNA-like structures to replicate in certain conditions. But many questions remain. Where would these conditions exist? Possibly in hot clays or underwater volcanoes where primitive life is seen. As yet, however, there is no widespread agreement about these possibilities, and the origin of the first living cells remains a mystery.

This is well summed up by the international science journal *Nature*. In recent years it has pursued an editorial policy that has not been sympathetic to any supernatural intervention in the origin of life or to Hoyle's panspermia ideas. However, it was forced to admit in 1994 that after many years of research into the origin of life,

> unfortunately, there is as yet not much that is tangible to report. . . . There have been many pointed investigations . . . but there is not yet an unambiguous pointer to the mechanism that may have led to the emergence of living organisms.[4]

So how do we get from primitive one-celled life to complex creatures? About 600 million years ago such complex creatures began to show up in the fossil record.

From Amoebas to Armadillos

The general view among biologists is that Darwinian evolution of organisms is fairly well understood. Evolutionary theory claims to explain the origin of complex life forms by small differences between individuals in a population having "survival value." Indeed the development of genetics in the twentieth century coupled with Darwin's ideas on natural selection have produced the "new synthesis," or "neo-Darwinism," which has many powerful advocates as an explanation of the emerging complexity of life.

Darwin's theory says that favorable variations in animals or plants tend to survive, and that over many generations these small variations lead to new species. It has been built on the evidence of

- the fossil record, which exhibits a progression from simple to complex structures
- similar structures in the anatomy of widely different species
- the modification of plants and animals by breeders
- vestigial organs like the "tail" of the human embryo
- changes due to geographical distribution
- the construction of an evolutionary tree through molecular biology

Genetics has provided the reason for the variations. As we have seen, genes are essentially DNA, and the genetic code is the sequence of the nucleotides that make up its long chain. Occasionally, when the DNA is copied, as when a cell divides, a small mistake occurs in the copying process. Alternately, radiation or poisonous chemicals can affect the DNA structure. These are mutations, most of which will be harmful, but a very few of which will be beneficial to survival. As the DNA code leads to the proteins, and the proteins control the entire chemical composition of any living creature, so changes happen that are then "selected" by the environment if they are beneficial to survival.

Nevertheless, this does not explain everything. Tim Hawthorne, professor of biochemistry at the University of Nottingham, writes,

> We know how DNA controls the making of proteins . . . but these are much the same in all living creatures. We need to know how growth and development are controlled, or why legs or brains or noses are a particular shape. How did the single fertilized egg from which each of us grew manage to produce all the different types of cells which go to make up skin and nerves, muscles and bones? Until we know more about the marvels of embryology we are not likely to understand the origin of species.[5]

Even with neo-Darwinism itself there remain some questions, many of which are the subject of a number of books by biologists, philosophers and those who argue from the standpoint of seven-day creationism.[6] Some of the following questions have been raised:

☐ Can Darwinian evolution be tested experimentally?

☐ Is it adequate to explain the development of new species or even new organs?

☐ Does the fossil record give enough evidence for evolution?

☐ Why have we found so few "missing links" in the fossil records?

☐ Is the mutation rate too slow to account for all changes?

In the face of such questions, some biologists have adapted the basic picture. For example, paleontologist Stephen Jay Gould, in the light of the fossil record's few "missing links" and the mutation rate's being too slow for all of the changes, suggests a "punctuated equilibrium" model with sudden, quite large-scale changes.

However, such questions have been robustly dismissed as problems for evolution by Richard Dawkins in a series of books and public appearances.[7] Dawkins does not see any of these questions as real problems. He predicts that random mutations plus natural selection can quantitatively meet the constraints of geological data in the time available.

Dawkins is not alone. One of the elder statesmen of biology, Ernst Mayr, professor emeritus of zoology at Harvard University, concluded a recent review of the subject with the words that "the basic Darwinian principles are more firmly established than ever."[8]

Dawkins's central thesis is that Darwinian selection is blind.

> Natural selection, the blind unconscious, automatic process which Darwin discovered, and which we now know is the explanation for the existence and apparently purposeful form of all life, has no purpose in mind. It has no mind, and no mind's eye. It does not plan for the future. It has no vision, no foresight, no sight at all. If it can be said to play the role of watchmaker in nature, it is the blind watchmaker.[9]

The divine Designer is dethroned by the blind watchmaker. That is, there is no innate drive to complex life, intelligence and consciousness. For Dawkins we are simply "gene survival machines." In a memorable phrase of the Nobel Prize-winning molecular biologist Jacques Monod, each individual evolutionary step is pure accident, "chance caught on the wing."

The implication is that complex life is highly improbable, not to be repeated elsewhere. It depends very highly on the right conditions and in the end is just a fluke. What a depressing conclusion for those who are attracted by the thought of alien life.

From Armadillos to Accountants

It is important to make clear the distinction between life and intelligent, self-conscious life. Before the claim of primitive life on Mars, one British betting firm had offered odds of five hundred to one of life being found elsewhere in the universe. It had offered such bets for the past twenty years and stood to lose £31 million. However, the bets were wisely worded to say, "NASA must confirm the existence of *intelligent* extraterrestrial life." The bacteria from Mars, if confirmed, would not qualify as intelligent. Nevertheless, the odds have now come down from five hundred to one to twenty-five to one!

The distinction is often confused. When people talk of other life in the universe, they really mean intelligent life, as they would want some communication with this life. Of course the discovery of any other kind of life in the universe would be clear evidence that the development of life would not be entirely random. By implication one could then say that the development of intelligence would not be entirely random either.

Some of this century's thinkers have gone further and pictured the inevitable development from amino acids to the single-celled organism to the complex living creatures and then to human-level intelligence. But that does not necessarily follow. Indeed, as I have already pointed out, life on Mars, if confirmed, did not evolve to the state of employing accountants to audit the cost of building all the canals!

The emergence of intelligence on Earth was dependent on such things as the onset of photosynthesis, the emergence of cells, the growth of multicellularity, the arrival of sex and the invasion of the land at the most basic level. This does not even touch things such as the development of a nervous system and other essential organs.

It is often argued that natural selection will lead to growing intelligence, because intelligence gives an advantage in the struggle for life. But this is too simplistic. John Barrow and Frank Tipler have argued the case strongly in terms of lineages, or strands of development in the evolutionary picture.

It is not intelligence alone which generates selective advantage; a

> sophisticated nervous system requires a huge number of support systems . . . to be effective. It is quite possible that no lineage on an earthlike planet will evolve the necessary support systems for a human level intellect, and possible that even if they do, the genetic coding of the support systems will be such that an increase in the complexity of the nervous system will be necessarily offset by the degeneration of some essential support organs in all the possible lineages on the earthlike planet.[10]

There are many science-fiction stories that postulate plants, reptiles and sea creatures with humanlike intelligence. However, on Earth there has been very little development in intelligence in these lineages in comparison with human beings. For example,

- Information processing in plants has developed so slowly that it would take many trillions of years to reach anything like the human capacity—much, much longer than the age of the universe, which is fifteen billion years.
- The reptilian metabolism is simply unable to support a large brain.
- The ratio of brain weight to body weight, which is thought to be a good measure of information processing or intelligence, seems to have developed to a certain degree in sea creatures and then stopped. This has been the case with dolphins, squid and octopus.
- It may be that intelligence is a hazard to survival rather than a help. A complex nervous system needs a longer time for gestation in the womb and then for development by teaching—times of great vulnerability and danger for the young.
- Even in primates a well-defined limit on the ratio of body weight to brain weight is reached in all lineages except that leading to *Homo sapiens*.

Intelligence does not seem to have a inevitable progression and does not seem to have survival advantage in its own right. Barrow and Tipler quote a leading evolutionist, C. O. Lovejoy:

> Man is not only a unique animal, but the end product of a com-

> pletely unique evolutionary pathway. . . . We find, then, that the evolution of cognition is the product of a variety of influences and preadaptive capacities, the absence of any one of which would have completely negated the process, and most of which are unique attributes of primates and/or hominids. Specific dietary shifts, bipedal locomotion, manual dexterity, control of differentiated muscles of facial expression, vocalization, intense social and parenting behavior (of specific kinds), keen stereoscopic vision, and even specialized forms of sexual behavior, all qualify as irreplaceable elements.[11]

There remains an enormous difference between humans and the higher apes, much of which is still a mystery. Michael Ruse comments, "Nothing yet . . . even scratches at an explanation of how a transformed ape could produce the magnificence of Beethoven's choral symphony."[12]

Depression increases for the believer in extraterrestrial intelligence. The view of virtually all leading biologists is that the evolution of intelligent life at a level comparable to human beings is so improbable that it is unlikely to have developed on any other planet in the entire universe.

The emergence of life does not necessarily imply the development of intelligence. If certain factors have to be just right for the first bacteria, then a whole series of other things need to be just right for the development to intelligence. Cells must combine to form viable bodies, and then nervous systems need to develop in complex bodies for the emergence of animal intelligence. It took about 250 million years before human intelligence emerged. In a universe teeming with life, we could still be alone.

Conscious of My Own Consciousness?

It is also important to hold a clear distinction between intelligence and self-consciousness. Some animals have conscious experience and limited mental ability, for example, chimpanzees can be trained to use sign language. Yet this comes nowhere near to the human ability for self-reflection. How does this develop?

If some of the earlier questions about the origin of life seemed complex, this is just as difficult. Indeed, it is a subject of great debate between scientists and philosophers, and has been for many years. What is the relationship between mind and brain? What do we actually mean by consciousness?

Opinion is split on whether we can understand consciousness. Daniel Dennett, one of the leading thinkers in this field, wrote a book titled *Consciousness Explained,*[13] although more recently he has been a little less confident, with the title *Kinds of Mind: Toward an Understanding of Consciousness*.[14] Dennett's view is that as we understand the neuroscience of the brain more, so we will more fully understand consciousness.

But this is not generally accepted. Philosopher Roger Scruton argues that Dennett describes only consciousness, not self-consciousness. There is a difference between possessing information and having an awareness of what is involved in possessing it. David Chalmers, professor of philosophy at the University of California, agrees. He believes that there is "an extra, irreducible ingredient" to self-consciousness.[15]

If this does not add immediately to the pessimistic scientific view for the believer in extraterrestrial intelligence, it does increase the complexity of the discussion. It is difficult to discuss whether aliens can be both intelligent and self-conscious if, concerning our own consciousness, as Stuart Sutherland puts it, "it is impossible to specify what it is, what it does, or how it evolved. Nothing worth reading has been written about it."[16]

From Accountants to Aliens

But as we noted at the beginning of chapter two, there are some dissenting voices, led mainly not by biologists but by physicists.

Physicist Paul Davies is one of the leading voices against the pessimism of the enormous improbability of life, intelligent or not, elsewhere in the universe. In his book *Are We Alone?* he pursues the view that if matter and energy have an "inbuilt" tendency to amplify and channel complexity, the odds are reduced dramatically of subsequent evolution of life and intelligence. Thus complexity can arise spontaneously through the process of self-organization.

Some would argue that although the probability of intelligent life is so small, if the universe is infinite, then all probabilities are bound to be fulfilled, so there must be other life. Davies rightly dismisses this argument on the grounds that if this is so, where does one stop? As he points out, an infinite universe means not only other life, but another "this life, this author, and this book"!

He suggests that the standard view of biology, that intelligent life is highly improbable, is built on two underlying assumptions. The first is the second law of thermodynamics, which was formalized in the work of Lord Kelvin and Rudolph Clausius in the nineteenth century. This recognizes that the amount of disorder (or entropy) in a closed system always increases. The general trend is that the universe is slowly dying. Taking this as the dominant worldview, evolution to complex life and intelligence contradicts it as a statistical fluke.

The second assumption is that since the controversy in the nineteenth century of intelligent design versus natural selection, biologists are extremely wary of any "guiding hand," whether it is divinity or even a law of nature that gives direction to a process.

Davies sees such a view as an intellectual copout. It does not address some fundamental issues such as the link, if any, between intelligence and the universe. Quantum physics suggests that it is the intervention of the observer's conscious mind that forms the link between the uncertainty of the quantum world and the certainty of the everyday world. Furthermore, why are the basic physical laws that permit such a complexity as life to develop at all special in their form, and why can we humans understand them?

Davies is very struck by these things.

> Consciousness and our ability to do mathematics is no mere accident, no trivial detail, no insignificant by-product of evolution that is piggy-backing on some other mundane property. It points to . . . the existence of a really deep relationship between minds that can do mathematics and the underlying laws of nature that produce them. We have a closed circle of consistency here; the laws of physics produce complex systems, and these complex systems lead to consciousness, which then produces mathematics, which can encode in a succinct and inspiring way the very underlying laws of physics that give rise to it. And we can then wonder why such simple mathematical laws nevertheless allow the emergence of precisely the sort of complexity that leads to minds—minds and mathematics—which can then encode those laws in a simple and elegant way. It's almost uncanny: it seems like a conspiracy.[17]

He then begins to resurrect a view similar to one proposed at the beginning of the twentieth century by philosophers such as Henri Bergson and William James, who believed in a force that represented the continuously creative nature of reality.

New advances in understanding chaos in physical systems are important for Davies. Chaos shows a link between randomness and order and the phenomenon of self-organization. It occurs in lasers, turbulent fluid eddies, chemical reactions and the formation of snowflakes. His argument is that if such spontaneous self-organization happens in physical systems, we should expect it in biological systems too.

He finds support in the ideas of Stuart Kauffman,[18] who combines biology and physics to suggest that there is an innate tendency of complex systems to exhibit order spontaneously. This tendency is used by natural selection and gets over the problems outlined above. The implication of this is that given the laws of physics, life will automatically emerge from an inert chemical soup under the right conditions.

This is not to say that there is some preordained goal, but simply that the trend from simplicity to complexity seems to be built into the laws of nature. In contrast to the second law of thermodynamics, there is some form of an organizing principle, an antientropy, which means that life does evolve regularly. According to this, the odds against the formation of life and subsequent evolution of intelligence are drastically lower.

Will There Be Alien Accountants?

This idea of antientropy remains hotly debated but has some supporters, such as the late Carl Sagan, astrophysicist and leading proponent of extraterrestrial life. Davies suggests that some biologists are beginning to prefer this option, although it is not a majority view.

Of course with only the example of life on Earth, it is impossible to be definite. However, it has become clear in recent years that many systems in nature in certain circumstances leap to states of greater complexity. It is as if greater complexity were preferred. It seems also that life on Earth arose very quickly, only a few hundred million years after the planet formed. Primitive life existed in a difficult environment consisting of volcanic eruptions, bombardment by asteroids and variable solar activity. The development of life on Earth certainly happened quickly and in difficult circumstances. Does this mean that life was an inevitable outcome of the laws of physics and chemistry under the right conditions?

This can be seen in two different ways. The self-organization might work in a general way depending on the conditions. When this complexity crosses a certain threshold, the system may be said to be living. There may be many ways that such organization can take place, and so there may be a vast variety of different life forms in a vast range of conditions. According to this view, life elsewhere in the universe may be very different from life on Earth, and the conditions may be not so important, for example, there may be no need for liquid water.

On the other hand, the processes could be much more focused in producing life very similar to that on Earth. This would demand very similar conditions to our own, such as liquid water and the correct temperature range.

Whichever way is correct, Sagan and Davies would expect the same general trend elsewhere, apart from Earth. Due to this built-in tendency, life would be beginning throughout the universe. Indeed, the existence of extraterrestrial life becomes a key test of this theory.

Auditing the Alien Proposal

What can be said of such a proposal? Davies bases it on three philosophical principles.

1. *The uniformity of nature.* The laws of nature are the same throughout the universe. This is a basic assumption of science. If the laws of nature were different in a different galaxy, then an understanding of the universe as a whole would be almost impossible.

2. *The principle of plenitude.* Whatever is possible in nature tends to become realized. Here Davies betrays that he is a physicist. He uses the example of how the discoveries of recent particle physics show that often, to our surprise, when particles are predicted due to mathematical possibilities, they are actually found.

This was a very popular philosophical assumption up until the middle of the nineteenth century. Its large-scale demise in the biological word occurred when it was realized that many species become extinct and give way to other species. Rather than all possible species existing together, some are just part of the evolutionary track to other species. Others simply die out with no apparent long-term survival.

The use of the principle of plenitude is somewhat controversial in the discussion of extraterrestrial intelligence. It can surely be used only if one already accepts some "law" of increasing complexity leading to intelligent life. The discoveries of particle physics are backed up by theories that rest on evidence and do predict the existence of these particles. It is not clear whether that is the case for intelligent life.

3. The Copernican principle. This states simply that we have no special place in the universe. In 1543 Copernicus published *On the Revolutions of the Heavenly Spheres,* in which he argued that the Sun, not Earth, was at the center of the solar system. (In fact in his calculations the planets rotate around the center of the Earth's orbit, which was not the same as the position of the Sun.) It was one of the key moments when Earth lost its special place as the center of everything.

This is accepted by most astronomers in the sense that Earth is not atypical in terms of its location in time and space or in terms of its laws of nature. Some, including Davies, go on from this to argue that this means there is nothing special about its biological circumstances either. But this is a matter of debate. Earth's circumstances may be just right in a number of ways to make possible the emergence of extraterrestrial life. Our very presence may be saying that our circumstances are special in some way. Recognizing this has been basic to an understanding of the anthropic principle, the realization that the laws of physics and the circumstances of Earth are very sensitively balanced to make life possible.

By joining these three principles together, Davies, with insights into chaos and the self-organization of some systems, argues for life elsewhere in the universe, although he is cautious about going the whole way to intelligent life.

It is interesting to observe that this "innate tendency in the laws of nature to bring forth life" has parallels within religious thought, in particular Christians such as American geologist Asa Gray, who sees God directing and causing the process of evolution. If it were shown that such a tendency to bring forth life were present in the laws of nature, wouldn't it raise the question of why this were so? A natural answer would be that God created a universe that has the inevitable consequence of life. Perhaps Davies's "conspiracy" is divine creativity.

Any Possibility of Life Out There?

However, we need to recognize that these claims, at the moment, are speculative. The last two of Davies's assumptions may be questioned when it comes to other life in the universe. The prevailing view among biologists is that life is very improbable indeed and intelligent life even more so.

Of course the key test for all of these theories is the existence of extraterrestrial life. The discovery of life elsewhere in the universe, especially if that life were intelligent, would overturn much of the standard evolutionary picture.

It is important to note that in all of these discussions it is assumed that life is based on the carbon atom. There is a great love in science-fiction circles of the idea that life elsewhere in the universe could be based on silicon, some exotic form of matter or even plasma. It is not at all clear whether such life is possible, and indeed such a claim can be a license for anything. Carbon is abundant in the cosmos, and we do know of at least one planet of carbon-based life forms, so it would seem reasonable to restrict our discussion at this point to such life. If we ever encounter a different form of life, then we can modify the assumption.

So how could we discover what types of other life could be out there? What are the claims that we have already encountered extraterrestrial life? It is to these subjects that we turn next.

4

Calling Occupants of Interplanetary Craft

Karen and Richard Carpenter were world famous for the kind of middle-of-the-road rock music you could buy elderly relatives for Christmas. So alongside "There's a Kind of Hush All over the World Tonight," it is quite a surprise to find a song about contact with extraterrestrial life. "Calling Occupants of Interplanetary Craft" came out as a single in 1977 under the subtitle "Recognized Anthem of World Contact Day." Its method of contact was somewhat interesting:

> In your mind you have capacities you know
> To telepath messages through the vast unknown
> Please close your eyes and concentrate with every thought you think
> Upon the recitation we're about to sing.
>
> Calling occupants of interplanetary craft . . .
>
> Together that's the way
> To send the message we declare—it's contact day.[1]

A cynical person would suggest that telepathy is probably the best way if that is how far English has fallen. Whether this experiment in telepathy worked or not we will come to later.

Nevertheless, this particular song is literally reaching out into the universe. The television signals that took it around the world seep out into space in an expanding circle, like a huge bubble swelling out from Earth at the speed of light. By 1996 a song played on the airwaves in 1977 will have traveled a distance of nineteen light years away from us. It would have reached the nearest star to the sun, Proxima Centauri, in 1981, although we do not know whether there are planets there, never mind intelligent life.

The planets around the star 51 Pegasi, which is forty-five light years away, will receive the Carpenters' message in the year 2022, and if there were any inhabitants there who would want to send us a song in return, we would receive it in the year 2067. Of course such inhabitants might already know of our existence, having heard by now the crackling noises of radio broadcasts since the early 1900s. Any inhabitants of Proxima Centauri would have learned a great deal more about us and as I write this might even be settling down to watch overnight coverage of the Olympics from Barcelona.

If alien civilizations are out there, they might not want to contact us after watching *Mork and Mindy* and *Dallas*. But one of the primary ways of finding out about other intelligent life in the universe is through indirect contact, by trading messages. We can either send them or look out for them. Perhaps messages are being beamed to us, or perhaps we can intercept an alien version of *Twinkle, Twinkle Little Star!*

Contact

Well known for his television series *Cosmos,* the late American astrophysicist Carl Sagan was an imaginative advocate for the existence of extraterrestrial life. In his 1985 novel *Contact*[2] (which was released as a movie in 1997), he speculates on how an extraterrestrial message might be received and the effects it would have.

His story concerns a radio telescope array that receives a signal consisting of a series of prime numbers. This is a key indication that intelligent life is sending the signal, as no astronomical object would be able to do this. Sagan cleverly weaves a number of themes into the story: the attention of the world media, the attitude of the government, what this contact means for religion, how careful one must be that such a message is not a hoax generated by someone on Earth. The message contains a number of different strands, one of which is instructions for a machine that will send a group of human beings across the vast distances of space.

Is all this just fantasy, or is it possible that this could happen? How seriously does the scientific community take the possibility of contact, either by messages or by space travel? For a short time in 1967 it seemed that something very similar to Sagan's story had happened.

Little Green Men in . . . Cambridge?

In 1965 Jocelyn Bell arrived in Cambridge as a research student to work with Anthony Hewish. Her Ph.D. work involved making a radio telescope. This was not the kind of radio telescope that was glamorous enough to star in the James Bond movie *Goldeneye,* like the Arecibo telescope in Puerto Rico.

Once the telescope was built, in a field a couple of miles away from Cambridge, Bell went each day to fill the inkwells and watch the paper chart churn out about one hundred feet of paper as the telescope measured radio signals from the sky that were passing overhead. One day in August 1967 she noticed a "ragged signal" that looked unusual and filled only a quarter of an inch in the hundreds of feet of chart. She checked back and saw the same signal a number of times in the earlier records. In November the team discovered that the signal was coming from a source emitting regular pulses of radio waves at intervals just greater than one second. They were able to calculate that its distance put it within our galaxy and that the object emitting the pulses was very small, no bigger than Earth.

A regular signal would be one of the signs that another civilization was trying to contact us. In fact Anthony Hewish had tentatively catalogued the source as LGM, short for "little green men"! Then the team was able to see that the source was not a planet orbiting a star. If this had been the case, they would have seen evidence of Doppler shift as it orbited. So if it was aliens, it was not coming from their home planet. Was it a spaceship or a radio beacon?

Bell checked the records and saw other such objects. Did this mean that there were lots of beacons or civilizations? In fact with further study it turned out to be a natural phenomenon. They had discovered the first pulsar, a spinning neutron star. When a star runs out of available hydrogen fuel, certain changes take place. In the case of stars larger than our sun, the star swells but then collapses due to gravity. This gravitational collapse is so extreme that at the core electrons and protons combine to form neutrons. The star's matter is forced into a sphere only a few miles in diameter. It is so dense that a teaspoon of this matter would weigh 100 million tons. Such objects can spin up to one billion times per second. The intense electrical and magnetic fields of these objects can produce a highly focused beam of radio waves, which appear from Earth to be short pulses because of its rotation. It is rather like the short pulses of light from a lighthouse.

Hewish received the Nobel Prize for discovering the pulsar, not little green men! This story highlights a problem that always dogs the search for extraterrestrial intelligence. How can we be sure that a signal is from an intelligent origin rather than from a natural phenomenon that at present we do not understand? We will see this again in claims of UFO sightings.

It's Aliens, Probably

The search for extraterrestrial intelligence (or SETI, as it's often called) works from the basic assumptions discussed in chapter three, that Earth is not a special place and that life has developed on other planets in the universe. However, in order for us to know that life is out there, with our present very modest abilities in space travel, further assumptions need to be added: such life would develop a level of intelligence that would allow it to transmit messages through space; and such life would want to communicate, would be as interested in finding us as we are with finding it.

Most discussions concerning the possibility of contact with other intelligent life feature the "Drake equation." This is a way of attempting to calculate the probability that a civilization will evolve in a star system and will eventually attempt interstellar communication. The Drake equation involves estimating

1. the probability that a given star system will have planets
2. the number of habitable planets in a solar system that has planets
3. the probability that life evolves on a habitable planet
4. the probability that intelligence evolves on a planet with life
5. the probability that an extraterrestrial intelligence will attempt interstellar communication within five billion years after a planet has formed

It is almost impossible to work out an answer to this equation. Only (1) and to a lesser degree (2) can be determined by experimental observation. The difficulty of all probability calculations is that they must be checked with a large sample. For (3), (4) and (5) we only have one case—Earth.

It is therefore not surprising that different people get wildly different results out of such an equation. American astronomer Frank Drake, after whom the equation is named, puts in his numbers and comes up with one thousand civilizations as advanced as our own in the galaxy. Others put in other numbers and come up with one advanced civilization in the galaxy, that is, us.

Hail Them on All Frequencies

Whatever the probability, the question then is how could we confirm the existence of extraterrestrial intelligence? There are at least 100 billion stars in our Milky Way galaxy and, more importantly, millions of different frequency bands that an alien intelligence might be using.

An example of the problem would be trying to tune a domestic radio into a specific radio station without any knowledge of its frequency or waveband. And what if you were doing it at a time when you did not recognize the program that the radio was broadcasting, or even more seriously, what if the station itself had closed down its transmission for a period? It would be a frustrating experience.

Magnify that problem many times and it becomes clear that simply pointing a telescope randomly at the sky is not the best way to start. Even with unlimited resources, which is never the case, it would be like looking for a needle in a haystack. Then there is the problem of recognizing the message. In the movie *Star Trek IV: The Voyage Home* Kirk and crew are confused by a message from an alien spacecraft directed at the Earth, until they realize that it is a message in whale song!

In 1959 two physicists at Cornell University, Giuseppe Coconi and Phillip Morrison, attempted to reduce the haystack. They suggested that if aliens were trying to communicate with us, then they would choose an obvious frequency to do it on. They further suggested that the frequency would be 1420 MHz, the frequency at which hydrogen, which composes over 70 percent of the matter of the universe, radiates radio energy. The frequency does not have to be exactly 1420 MHz but could be a multiple of this frequency, but at least it is somewhere to start. It is the same principle as TV advertising: put your message on the channel you know people will be watching.

Frank Drake began the search by observing two nearby stars at this radio frequency. Nothing was found, but the possibility of finding an alien message suddenly became scientifically feasible. Since then there have been numerous small-scale attempts to detect signals from extraterrestrial civilizations, and radio astronomers have searched large sections of the sky with no success. The only exception has been one unexplained signal in 1977. It was called the "Wow" signal after the startled researcher at Ohio State University scribbled the word on a printout. This signal has not been seen again, and to some it has not been satisfactorily explained.

Even when reducing the size of the haystack, the SETI challenge remains formidable. Astronomer David Hughes writes,

> First you have to point your radio telescope in the direction of a star that might be the parent of a planetary system. . . . Then for each star you have to search a radio window that stretches from 1 to 10 GHz and contains 100,000 million 0.1Hz bandwidth channels. No wonder you are thankful for your computer's Fourier-transform superprocessor; no wonder you are worried about the fluctuations in the background noise resembling an artificial signal. And even overlooking the fact that your search might last for the lifetimes of many generations of scientists, you still have to contend with the fickle nature of scientific funding agencies who are only too happy at times, to suggest that you are wasting your time and their money![3]

Bagging Little Green Fellows

In October 1992 a new phase began with great optimism. NASA began a ten-year, $100 million search for extraterrestrial intelligence. It was launched on the five hundredth anniversary of the arrival of Christopher Columbus in America. Originally called Project Columbus, it quickly became Project Phoenix. It attempted to pick up signals from one thousand nearby stars similar to our sun, out as far as one hundred light years. At the same time it also attempted to scan further afield for a message. If aliens lived around the nearest stars, then we might pick up their signals in the same way that they would pick up our TV signals. Or if aliens were colonizing the galaxy, we might pick up some of their radio communication between the stars.

But such optimism quickly met harsh economic realities. In 1993 the NASA program was closed down by Congress. Senator Richard Bryan, one of its leading opponents, called it a "great Martian chase," "a waste of taxpayers' money," and pointed out that "we have yet to bag a single little green fellow."

Since then the search has been carried out by privately financed groups. Frank Drake, the founding father of SETI, heads the SETI Institute in California. It was set up as a distinct organization in 1984 and is financed by millions of dollars of private donations. It needs around $4 million a year and has some high-profile backers, including Paul Allen of Microsoft, Gordon Moore of Intel, and William Hewlett and David Packard of Hewlett-Packard.

In 1995 the SETI Institute sponsored Project Phoenix, using the world's largest radio telescopes. It can analyze more data than all the previous SETI efforts put together in just a few minutes. It is 100 trillion times more sensitive than Drake's original search. It uses the U.S. National Radio Observatory in West Virginia and monitors 28 million channels at once, compared to Drake's one channel. The project's first step is to study a thousand nearby stars similar to our sun.

Another private group is the Planetary Society. This was headed by the late Carl Sagan and receives sizable donations, such as $100,000 from Steven Spielberg, the producer of *ET.* One of its projects is the BETA telescope near Boston, Massachusetts. BETA is the acronym for Billion-channel Extraterrestrial Array, and it consists of a dish-shaped antenna that is eighty-four feet across. Each day, as the Earth turns, it sweeps a circular swath across the sky, elevated at a slightly different angle from the horizon with each successive turn. It receives radio waves and analyzes them through a supercomputer that attempts to filter out cosmic radio noise. Every two seconds BETA captures enough data to fill a CD-ROM, about 22 million megabytes of data per day.

This amount of information is totally beyond the capacity of the human mind to store or interpret. The young research student Jocelyn Bell would not have time to fill up the inkwells, never mind look at the data to try and see "ragged signals." So within the BETA's computer is a program that decides if a signal is artificial. If such a signal is received, the telescope abandons its search of the skies and fixes on this one point for further analysis.

Yet another project, known as SERENDIP (Search for ExtraTerrestrial Radio Emissions from Nearby Developed Intelligent Populations), is based at the University of California and chaired by science-fiction author Arthur C. Clarke. It has studied the skies using the world's biggest telescope, the one-thousand-foot dish at Arecibo Observatory in Puerto Rico.

Message in a Bottle

Because of the large amount of time and money spent on these projects, the question is raised as to what we are actually looking for. An artificial signal would alert us to the existence of other intelligent life, but it may not be direct contact. It might simply be a radio beacon. Of course the beacon might carry information about the life form that made it. In a sense it would be a cosmic message in a bottle.

Pioneer 10, the first humanmade object to leave the solar system, carried a plaque with basic data about human beings and a picture of a man and a woman. The Voyager spacecraft that explored the outer planets carried more information as they headed off into interstellar space.

When discussing receiving or sending a message, we must reemphasize the difficulty involved in communication. In November 1974 the strongest humanmade signal ever transmitted was beamed by radio telescope in the direction of the Great Cluster of the constellation Hercules. Some people objected to this being done in case it revealed our position to aliens who might then come and conquer us!

The problem is, however, that such a message may literally have to travel for millions of years before it is received by an intelligent civilization. If the message is understood, a reply could again take millions of years to return. This does not lead to a very stimulating conversation.

The problem of delay in communication is a very important one. Unless the galaxy is teeming with intelligent life in the way that *Star Trek* envisages, it could be that intelligent life is there, but we may not know about it. Or it may be that by the time we would receive a message, the alien civilization would have gone out of existence. The universe could be full of life that was unable to communicate, and we would never know.

A message can reach its destination only at the speed of its method of transportation. Einstein's theories of relativity indicate that electromagnetic radiation (of which light, radio waves and microwaves are a part) travels with a constant speed everywhere in the universe. Astronomers talk about distances in terms of how far electromagnetic radiation can travel in a year, which is a light year. This is a natural limit to how quickly messages can be sent.

The very nearest stars are a few light years away, and the nearest galaxies are millions of light years away, so there have been suggestions for forms of communication that might be faster. In science fiction this often takes the form of "subspace" communication, or in physics the talk of faster-than-light particles or some of the interesting effects of quantum mechanics. All are extremely speculative, and at the moment there seems to be no sure way around this problem.

A Vast Loneliness

"More loneliness / than any man could bear / Rescue me before I fall into despair." In the early 1980s the Police sang about loneliness in the song "Message in a Bottle." The lyrics are not inappropriate to the loneliness that SETI has so far revealed.

Although in one sense the problems involved in receiving a message are huge, it is not insignificant that we have so far heard nothing. As we have seen, there are technical reasons for doubting that these searches will be successful. But there are a lot of stars out there and, if the advocates of extraterrestrial life are to be believed, many civilizations as well.

In a paper published in the *Quarterly Journal of the Royal Astronomical Society* in 1992, George Lake[4] quotes a private communication from Carl Sagan stating that the absence of evidence for extraterrestrial intelligence from the 10^{20} extragalactic stars that have been surveyed up to this date is already a remarkable result. Lake goes on to argue that this is due to not the lack of extraterrestrial intelligence but to the fact that the time scale for the evolution of an advanced civilization is short compared to the time it takes for any message to cross the vast distances between the galaxies. We are effectively surveying many galaxies at a time before civilizations have evolved.

There are other attempts to explain the absence of any signal from extraterrestrial intelligence. P. Wesson[5] suggests that if intelligent life is sparse in the universe, it might well lie beyond the cosmic particle horizon, the region of space that is causally connected to us. As we observe the universe, there exists a boundary corresponding to the age of the universe beyond which we cannot see. That is, if the universe is fifteen billion years old, we can see only regions of space where light has traveled for fifteen billion years to get to us. If it takes a civilization four billion years to evolve, as it has on Earth, the region that can contact us shrinks to eleven billion light years. Reviewing the high odds against the evolution of intelligent life, Wesson concludes that we are alone in the observable universe. Extraterrestrial life may exist in the regions we cannot see, but it cannot be contacted.

Property of the Klingon Empire?

Yet other people cling to the ideas that we are not looking in the right way or that aliens are holding back contact until we are ready for it. Ukrainian scientist A. V. Arkhipov suggests we should be looking for alien artifacts. If there were advanced alien civilizations constructing technology in space, there would be, as a result of such things as accidental explosions of satellites, a leakage of alien "artifacts" from their solar system into the rest of the galaxy, some of which might eventually fall to Earth. Arkhipov calculates that Earth might have accumulated about four thousand artifacts of around one hundred grams throughout its history. This seems to be a somewhat high estimate for a number of reasons, not least because he assumes that 1 percent of planetary systems are manufacturing artifacts. This puts the number of advanced civilizations in our galaxy at hundreds of millions. And there is the further problem of how we would recognize whether an artifact was alien. It is unlikely that it would have "Property of the Klingon Empire" stamped on it in English.

The existence of alien artifacts was the center of one of the most famous science-fiction novels and films. In Arthur C. Clarke's *2001: A Space Odyssey* astronauts on the moon discover a strange black obelisk. The implication is that it was placed there to be discovered only when human beings were advanced enough to see it. Frank Drake picks this theme up in terms of the SETI project.

> If we want to join the community of advanced civilizations we must work hard as they must. Perhaps they will send a signal that can be detected only if we put in as much effort into receiving it as they put into transmitting it.[6]

We have here begun to tread into the realms of the alien mind. What would be their motives for communicating, and how would they relate to us? To this we will return shortly, but before we do we need to go a little further along the direction of communication with extraterrestrial intelligence and ask: Instead of radio communication, is there any possibility of direct contact by space travel?

5

To Boldly Go Where No Alien Has Gone Before

It all seems so reasonable in the science-fiction world. After all, radio communication that takes millions of years does not make riveting reading or movies. For that we need face-to-face contact and thus a method of transport across the vast distances of space. The problem of space travel has been a perennial problem for science-fiction writers. Ever since Albert Einstein pointed out that no spaceship could be accelerated faster than the speed of light, an upper limit has been set that makes the vast majority of space journeys between the stars totally impractical.

Some authors have chosen to avoid the problem altogether. Others have developed impressive-sounding concepts such as hyperspace to transport heroes over the vast distances. In this space race the USS *Enterprise* stands supreme. As long as the dilithium crystals are all right (and one wonders why they never carry an extra set!), the captain engages warp drive and the ship speeds off over incredible distances within the fifty minutes of each episode. Is such a concept possible?

Of course, for all the impressive science terminology, it is fiction. After all, the USS *Enterprise D* featured in *Star Trek: The Next*

Generation has an arboretum, laboratories, an education center and a number of virtual-reality holodecks, but only two bathrooms for a crew of 1,012! In fairness to its creators, "warp drive" has been given some scientific basis. It is imagined that the warp drive contracts the space-time in front of the *Enterprise* and expands it behind, making it possible to travel faster than the speed of light.

This consideration of interstellar transport is important in two areas. In the next chapter we will look at possible evidence that our planet has been visited by aliens. Such evidence must be viewed in light of the chances of UFOs actually getting here. Second, and perhaps more important, space travel is part of one of the most powerful arguments against the existence of extraterrestrial intelligence. It is to this that we turn first.

If They Existed, They Would Be Here

At a luncheon in 1950 physicist Enrico Fermi devised what is often called Fermi's paradox or the space-travel argument against the existence of extraterrestrial life. He put it simply: "If they existed, they would be here."

He argued that if Earth is not unique in having intelligent life, then civilizations should already have evolved billions of times in the galaxy, since there are billions of stars older than the sun. If any one of these civilizations wanted to colonize the galaxy, they could have done so by now, even using technology that is almost within humanity's grasp. But there is no compelling evidence that any aliens have visited Earth, much less colonized it, so we must conclude that we are alone in the galaxy.

This is quite a powerful argument. The emergence of humanity relative to the age of the universe is really quite late. If the age of the universe were represented by the whole of the *Encyclopaedia Britannica,* then humans would appear in the last sentence in the last paragraph on the last page.

If other life were plentiful in the universe, then we would expect it to have developed before the emergence of intelligent life on planet Earth. One can then calculate how long it would take an intelligent civilization to colonize the Milky Way galaxy. So if another intelligent civilization were out there, it would already be in our neck of the woods.

Of course there are those who would object immediately to the statement that there is no compelling evidence that aliens have visited Earth. We will come to that in the next chapter. But let us for the moment look at the question of space travel in a little more detail.

From the Pacific to the Pleiades

Sending human beings on a spaceship would be very difficult. The energy costs alone of maintaining an environment on board would be very high. In addition there is the problem of the vast time scale for such a journey. One person might make it to the nearest star, but in order to go further, new generations would have to be born and raised along the way.

The other alternative, a favorite of science fiction, is to deep-freeze human passengers and wake them up when they arrive. This would reduce the energy needed during the journey, but whether or not it is scientifically possible is still a very open question. Of course, while in deep freeze the passengers could be eaten by aliens, switched off by mad scientists, or drift for ever because the guidance computer has a bug!

An alternative method for galactic colonization would be to send machines rather than people. This takes two forms. The first is to send a spaceship carrying either human fertilized eggs or (if it were possible) the building blocks and instructions on how to genetically create humans. When the spaceship finds an appropriate planet, new humans could be "born" and "raised" by the ship's computers and robots.

The second form is simply to dispense with humans. This would involve sending space probes that could collect and send back data while being able to make new space probes when they encounter the right raw elements in planetary systems. This is called a von Neumann probe after the scientist who suggested it was possible.[1] It is a self-replicating universal constructor with intelligence comparable to the human level. It could be instructed to conduct scientific research and transmit back the results, while at the same time to search out construction materials to make several copies of itself, which could be sent on to other star systems. It is generally thought that such a machine could be developed within a century. When combined with better rocket technology, which is feasible, such machines could travel at up to one-tenth the speed of light.

Astrophysicist Frank Tipler, one of the strongest advocates of the space-travel argument, has suggested that the island hopping of the South Sea Islanders across the Pacific Ocean is a good model for this. They arrived at an island and established a colony. After they spent some time there to allow the population to grow, a new expedition was sent to another island, and so the process continued.

In a similar way, "planet hopping" is limited only by the time it takes to get from one planet to another (that is, the speed of light) plus the time spent on a planet to get going to another. The speed of light would take one across the Milky Way in 100,000 years, and the time needed on a planet to prepare for the next step would be much shorter than this. So the time needed to colonize across the galaxy, even if one is not traveling at the speed of light, is very short compared to the age of the galaxy, which is not less than ten billion years old.

Tipler estimates that an extraterrestrial intelligence could explore or colonize the galaxy in less than 300 million years and even as fast as one million.[2] But some have argued against Tipler's calculations. Tipler assumes what is called a "free expansion model" where essentially all the probes are sent to new stars. Newman and Sagan[3] apply a "diffusion model," which takes into account forward and backward motion; that is, some probes may be launched to systems where there are already probes. However, this increases the colonization time by a factor of only three. Thus even with the longest possible estimate of colonization time, aliens should be in our neck of the woods within one billion years. This is still at most just one-tenth of the age of the galaxy.

Therefore, the argument concludes that the absence of any extraterrestrial intelligence in our solar system means that such space-traveling aliens apparently do not exist and have never existed in our galaxy.

ET, Stay Home

There are only two ways to get out of this conclusion. One is to maintain that aliens *are* here. Another is to argue that some factor makes galactic colonization extremely unlikely.

Paul Davies questions whether it is feasible to build a von Neumann probe over such a time scale and also whether it would be economically viable. He then goes on to ask whether an alien race would launch such a program of galactic colonization. Indeed, although we might talk of doing such a thing, economic constraints mean that we have only briefly surveyed the planets in our own solar system. And we often take the view that if aliens are out there, they will contact us. An extraterrestrial intelligence may be thinking the same thing about us.

Newman and Sagan suggest it is almost impossible to know whether extraterrestrial intelligence would or would not be motivated toward colonization. We cannot presume that their social structures, ethics and aspects of culture would be similar to ours.

Barrow and Tipler are not convinced by such claims. They counter with several points.

1. If another civilization were trying to use radio contact, then there is no reason why space colonization should not be seen as a better way of contact. There are many advantages to colonization, and so to argue

against colonization is to argue against any contact at all. They are out there, but they are so silent that we cannot know that they are there.

2. The behavior pattern not just of human beings but of all other living things on our planet suggests that expansion into new environments or colonization is basic to all life, not just intelligent life.

3. By colonizing the stars a civilization increases the probability that it will escape the death of its own star when that star runs out of its available hydrogen fuel.

4. Any fear of von Neumann probes getting out of control is unlikely.

So if there are no strong reasons for extraterrestrial civilizations to stay at home, in order to get around the space-travel argument we are left with the option that aliens are here. Perhaps they are here but we cannot detect them.

The Zoo Hypothesis

Perhaps our planet has already been colonized. Earlier we discussed the claims that Earth was seeded for life by an extraterrestrial intelligence. Perhaps there is a von Neumann probe already in the solar system and we have not detected it. Maybe such a probe has been hidden from us. We are being watched and studied without knowing anything about it.

This is the so-called zoo hypothesis.[4] It gets around the space-travel argument while maintaining that there is no real evidence in the solar system for probes or aliens from another civilization.

We might well argue that an advanced civilization would have the technology to make this possible. The trouble is, however, whether it is really possible to keep such a secret. Why would an alien civilization want to keep us in the dark? Surely they would note the differences between our levels of technology and would see that we would not be a threat to them. And what about the possibility of an accident, which has often been the stumbling block to top-secret technologies on

Earth? Further, would a galacticwide civilization be able to police all of its own beings? Perhaps a message would be sent by a group who believed that contact with Earth was important—an alien leak!

The zoo hypothesis does not seem to be totally convincing. But before we leave, we need to wander into the realm of faster-than-light travel.

Warp Factor Six, Engage!

So far in this discussion we have assumed that the speed of light is the upper limit to any space travel. Science-fiction writers have disagreed, and in recent years some scientists have suggested ways in which faster travel might be possible.

The difficulty with faster-than-light travel is Einstein's theory of special relativity. As a spaceship accelerates toward the speed of light, it actually increases in mass. We do not see this effect in everyday life because the speeds that we experience are so much less than the speed of light. At such speeds the increase in mass is imperceptibly small. However, as a spaceship approaches the speed of light, the increase in mass means that it needs more energy to increase its speed. At the speed of light the amount of energy needed to accelerate the spaceship becomes infinite, and so this forms an upper limit to how quickly one can travel between the stars.

So a journey to the nearest star would take a spaceship that was traveling close to the speed of light more than four years. Technical and economic limitations mean that the ship would never be traveling at more than a fraction of the speed of light, and so the travel time would be much longer.

Of course, this is the travel time as seen from Earth. One of the other curious things about Einstein's theory is that time runs slowly if one is traveling very close to the speed of light. Therefore, if a person boarded a spaceship that accelerated close to the speed of light, then that person's measurement of time would be very different from that measured by mission control on Earth. It would appear to the traveler that the journey was taking much less time than it would appear from mission control.

Does that mean that astronauts would not have to go into a deep freeze for long journeys? Not really. It is only close to the speed of light that the time-dilation effect, as it is called, really matters. For example, the average velocity of the Apollo mission to the moon was about one-millionth of the speed of light. At this speed, which is still very high by everyday standards, time dilation would not be noticeable. It would take astronauts four million years to travel to the nearest star and forty-five million years to travel to the star 51 Pegasi, where we have seen planets. And that is barely touching the galaxy. It is not even out of our back yard.

If we were able to accelerate a spaceship to one-tenth the speed of light, which is believed to be theoretically possible within a century, then the journey to the nearest star would take forty years as measured by mission control, but the astronauts would measure a time some seventy-three days shorter. A journey at this speed would take 450 years to reach 51 Pegasi, but the astronauts would measure only 448 years. A saving of two years is not a lot to write home about, if there could even be a way of sending a speedy message back.

Even at half of the speed of light, which is extremely optimistic, it takes eight years to reach the nearest star with astronauts "saving" one year and twenty-six days, while it still takes ninety years to 51 Pegasi with astronauts arriving in their own time of seventy-eight years. Still not a great saving on such a journey.

We can see from such figures that the speed of light is an unfortunate barrier to realistic space travel between the stars.

You Canna Change the Laws of Physics, Captain!

However, in the last few years scientists have been taking more seriously the possibility of some kind of warp drive that is faster than light travel, or at least faster than light communication.

We do know of at least one phenomenon that seems to communicate information faster then the speed of light. In 1935 Albert Einstein with collaborators Boris Podolsky and Nathan Rosen highlighted what they believed was an unacceptable consequences of quantum theory. In what is now called the EPR experiment or paradox, they pointed out that once two quantum particles such as electrons have interacted with each other, they retain the ability to influence each other even though they are separated by extremely large distances. For example, imagine two electrons that interact with each other. If I examine one of them, this has an instantaneous effect on the other, even if it is at a great distance, such as on the other side of the galaxy. Einstein felt this showed that quantum theory was incomplete.

Although this seems to go against everything we assume about the world, observations have confirmed that this really happens. Einstein was wrong, and quantum theory is right. Does this suggest faster-than-light propagation of information is a possibility? Certainly a message cannot be sent from one electron to the other at the speed of light, as the change is instantaneous. What the EPR experiment is demonstrating, at the quantum level, is that at the level of the particles that make up atoms, there is, in John Polkinghorne's phrase, "togetherness in separation." It does not contradict the special theory of relativity, it just says that at the quantum level the speed of light is not such a barrier.

Another theoretically possible way of sending information faster than the speed of light is by particles called *tachyons*. No one has ever seen a tachyon, and their existence remains controversial, but within Einstein's theory they can exist. The special theory of relativity says that a particle cannot be accelerated from a speed below the speed of light to a speed above it. But if someone creates a particle already traveling faster than the speed of light, then the theory does not rule out such faster-than-light travel. This could be used for communication as long as such tachyon particles could be created at one end and then detected at the other end.

There are other interesting speculations concerning travel as well. When preparing his novel *Contact,* Carl Sagan asked some colleagues whether or not it was possible for space travelers to cross the vast distances of the universe by means of "wormholes." Since the 1930s it has been known that the equations of general relativity allowed the possibility of very small "tunnels" linking one black hole with another somewhere else in the universe. Sagan's colleagues found that under special circumstances such wormholes could allow the possibility of travel. A ship could enter a black hole in one part of the universe and emerge elsewhere, a sort of shortcut. The main problem with this (apart from getting anyone silly enough to try it for the first time) is that wormholes are unstable. Any attempt to pass anything through would cause the wormhole to collapse.

Another method of travel could be warp drive. There have been recent suggestions of distorting space-time ahead of and behind a spaceship in such a way as to permit it to travel at faster-than-light velocities as measured by observers outside the distorted region. However, how possible this would be, given our technological limitations, is not certain. We will have to wait for Scotty from *Star Trek* to be born—although the majority of scientists would bet that if faced with such a problem, Scotty would realistically reply, "You canna change the laws of physics, Captain!"

They're Already Here

If faster-than-light travel or communication proved to be possible, what would be the implications? It would reinforce the space-travel argument against the existence of extraterrestrial intelligence in the galaxy and extend it to the universe. The time needed for galactic colonization would drop to perhaps one thousand years, which would perhaps allow the maintenance of social coherence and central political control in such a colonizing civilization such as Isaac Asimov's Galactic Empires or Gene Roddenberry's Federation. This may help in operating a policy of noninterference such as the zoo hypothesis, but it actually leads to many more spaceships arriving in the solar system.

Although there are uncertainties, the space-travel argument seems to be a powerful one against extraterrestrial intelligence, at least in our galaxy. This does not rule out extraterrestrial intelligence in other galaxies, but there the problems of travel and communication become extremely important. Messages would take millions of years to exchange unless some form of faster-than-light communication were found.

However, the space-travel argument works only if there is no evidence of aliens in the solar system. In the discussion mentioned at the beginning of this chapter, Hungarian physicist Leo Szilard replied to Enrico Fermi's "If they existed, they would be here" with, "They're among us and are known as Hungarians!"

There are increasing numbers of people in the world today who argue that aliens are among us. Much of the mainstream scientific community disregards talk of UFOs, alien abductions and direct contact, but this is a mistake. Such claims need to be investigated, even if answers may prove elusive.

6

Crop Circles, UFOs, Abductions and All That

Reg Presley is an Englishman who found fame as the lead singer of the rock group The Troggs. His song "Love Is All Around" topped the British charts for the group Wet Wet Wet and was part of the soundtrack for the film *Four Weddings and a Funeral.*

Presley doesn't live the stereotypical rock-star lifestyle, blowing his money on extravagances, but instead he uses his royalties to research crop circles. He believes that these circles and designs found in fields of corn, which have been appearing since the 1980s, are formed by aliens, who are either warning us of ecological disaster or refueling on their long interstellar journeys.

Although some have claimed that crop circles are not a new phenomenon, Britain in the late 1980s witnessed a spate of crop- circle appearances. By the early nineties these circles had expanded into spirals, triangles and lines of ever-increasing complexity. Besides the alien explanation, many other theories were suggested for their cause, from the downwash of helicopters to complex symbols of Earth's distress at pollution to hundreds of hedgehogs stamping around in circles.

One enthusiast, Terence Meaden, claimed that vortices of air were responsible. But he was set up on camera by a British TV station. They created their own circle and then filmed Meaden as he pronounced it genuine and was then told it was not.

In 1991 two artists, Doug Bower and Dave Chorley, claimed they had created dozens of circles since the 1970s using only ropes and planks of wood. Most circles are now accepted as hoaxes, although some fluid-mechanics experts have suggested that some are genuine, the result of vortices of air being swept off steep slopes and blowing on fields downwind.

It should have been realized that these were hoaxes when the circles appeared only in Britain. Out of all the open spaces on the planet, were only cornfields in England suitable for refueling spacecraft that had traveled thousands of light years?

Yet there are some, Presley included, who still believe the alien theory, so much so that a recent book *(Round in Circles)*[1] written by an American reporter was denounced as part of CIA coverup propaganda. Supposedly the CIA knew that aliens were the source of crop circles, so they employed the author to write a critical account of such ideas to mislead the public.

The belief that aliens make crop circles is at the fringe of the "UFO community," which has become big business in recent years, yet it illustrates some common themes that occur in many aspects of this area of study: unexplained phenomena, the reality of hoaxes and the belief in official coverups.

We will see these themes recurring as we look at the question of whether there has been direct contact with aliens through the sighting of UFOs, the capture of alien spacecraft or aliens themselves, or the abduction of humans by aliens.

It's a Bird, It's a Plane . . . or Is It an Alien?

Kenneth Arnold coined the term *flying saucer* in 1947 after seeing an object in the skies over Washington state. He was flying over the Cascade mountains hoping to pick up the $5,000 reward the government had offered for sighting a crashed transport plane. As he flew he saw nine disk-shaped objects traveling at incredible speed. He said to reporters later that they moved like "a saucer skipping over water." A reporter suggested "flying saucer," and the name stuck.

The sightings went on. Within a month the air force had received 850 UFO reports. Due to movies, cold-war fears and the MacCarthyite paranoia in America, people were looking at the skies and seeing strange objects. But it was not as if this belief in aliens visiting the planet were something totally new. Between 1945 and 1947 Raymond Palmer of the science-fiction magazine *Amazing Stories* had boosted his circulation to 250,000 with stories about space aliens he presented as fact. Interestingly enough for what was to happen later, these aliens also kidnapped humans.

By the late 1950s tales of meeting with and being abducted by aliens began. George Adamski was probably the first to claim that he had been contacted by aliens. He produced some stunning photographs of flying saucers and wrote two bestselling books, *Flying Saucers Have Landed* and *Inside the Spaceships.*

From the titles of the books it is clear that Adamski claimed something more than just seeing flying saucers. He claimed that he was taken on a trip to the moon by blond female Venusians (who no doubt, as we have seen in chapter two, were thick-skinned enough to survive the sulfuric acid) and that an alien leader revealed to him that they were here to save the world from nuclear radiation. However, Adamski was eventually discredited. Rejected as a writer of fiction, he had sold his books by presenting his stories as fact.

Nevertheless, the study of UFOs was growing in response to the many claims. J. Allen Hynek, who founded the Center for UFO Studies in 1973, had several classifications for contact.

1. Close encounters of the first kind. This was the observation of any UFO within about five hundred feet of a witness. There were many of these reports from the late 1940s onward.

2. *Close encounters of the second kind.* This involved trace evidence of UFOs, such as leaving marks on the ground, interference with engines or radios, or frightening or mutilating animals.

3. *Close encounters of the third kind.* This was the observation of the occupants of the UFO, and it was made famous as the title of Steven Spielberg's 1977 film.

In the 1970s Billy Meier, a Swiss farmer, claimed a series of contacts with visitors from the Pleiades, complete with an astonishing catalog of UFO photographs. Other popular accounts began to appear that claimed UFO sightings had been reported down through the centuries. However, many of the medieval accounts repeated in book after book are actually hoaxes. And it seems that more recently there has been an evolution in alien craft from saucers to triangles.

It is claimed that three million people in Britain have seen a UFO. Many of these claimed sightings of the first kind can often be explained easily. Even those who believe in the existence of alien spacecraft acknowledge that at least 95 percent of UFO sightings can be explained in terms of airplanes, aircraft lights, meteors, satellites, searchlights, flocks of birds and laser light shows at rock concerts, as well as the possibility of hoaxes.

The planet Venus is probably the most often seen UFO. More and more people are unused to seeing natural phenomena in the sky, in large part due to the effect of street lighting, which brightens much of the night sky. As this trend increases, some natural phenomena like the bright planet Venus can be easily mistaken for a UFO. Some people point out that many sightings of UFOs cluster around geological faults. Are there unusual geological effects that lead to strange atmospheric effects?

Sometimes the reported sightings are blown out of proportion to justify that UFOs are spaceships. In November 1996 an amateur astronomer, Chuck Shramek, claimed that he had photographed an object following in Comet Hale-Bopp's wake. This connection was pushed widely on the Internet, with photographs apparently showing the UFO. The Heaven's Gate cult took this to be their "suicide" spaceship. In fact, it was nothing more than a background star.

There are reported incidents where strange lights or objects are sometimes accompanied by radar traces. Of course technology that is very much humanmade can also be mistaken for alien craft. In December 1978 a Soviet booster rocket entered the atmosphere over Europe. This led to a spate of reliable witnesses claiming to have seen a UFO with light coming out of its portholes that was about to crash. The American stealth bomber, a strange, triangular-shaped black airplane, was often mistaken for a UFO before its existence was acknowledged. It is reasonable to suppose that other secret projects may have been mistaken for aliens. This might explain why sightings often cluster around military bases.

Yet some sightings do remain unexplained. But before we bring up government conspiracy theories, some governments have given some attention to UFOs. The British government's defense ministry has a two-person Secretariat (Air Staff) 2a, which deals with a number of issues of public concern, of which UFOs are just a small part. Even so, they deal with between two and three hundred reports of unexplained aerial sightings a year. The U.S. Air Force's Project Sign investigation in 1948 concluded that 20 percent of cases were inexplicable. However, the U.S. government's Project Blue Book investigation, which ran from 1952 to 1969, concluded there was nothing to worry about.

It would be much easier if the spacecraft were as large and as public as they were in the movie *Independence Day,* where huge flying saucers positioned themselves conveniently over the major cities of the world. But purported alien craft seem to come in all shapes and sizes and choose to reveal themselves in very odd places. They have an elusiveness that is both attractive and frustrating. If only we could get our hands on one! But then again, some claim that we already have.

Just What Happened at Roswell?

This most celebrated story contains all the classic ingredients of a UFO sighting. It took place on July 4, 1947, near Roswell, New Mexico. Apparently an alien spacecraft crashed on ranch land during a thunderstorm, not far from an air force base. The London *Times* reported that the army announced on July 8 that they had found an object at this location that resembled a "flying disk," and it had been sent to the military research center at Wright Field, Ohio. The statement said, "The many rumors regarding the flying disk became a reality yesterday. . . . The flying object landed on a ranch near Roswell some time last week."

The land now belongs to Hub and Sheila Corn, and for fifteen dollars tourists can see the exact place where the aliens crashed. W. W. "Mac" Brazel, a local farmer, heard the impact above the sound of the loud storm. When he went out to investigate the next morning, he found a trail of debris about three quarters of a mile long and two hundred yards wide. It consisted of shining foil, fibers and shards of a tough rubbery substance that had odd markings on it. Brazel later reported that the debris weighed no more than five pounds.

On July 9 the army retracted its statement and claimed that the wreckage was a weather balloon. At that time UFOs were being reported daily, so at this stage there was not a great deal of immediate interest.

In fact not much was to happen until 1978. A "UFOlogist," Stanton Friedman, met Major Jesse Marcel, the intelligence officer who had collected the debris with Brazel and whose name appeared on the initial press release. Marcel believed the wreckage to be extraterrestrial and that a coverup had been ordered by his superiors. Friedman interviewed other "witnesses," and in 1982 his research was published in *The Roswell Incident* (with Charles Berlitz and William Moore).[2]

Stories began to develop. It was claimed that the authorities had discovered the dead alien crew at another crash site and had taken them to the Fort Worth air base for postmortems. And some three hundred witnesses came forward talking of "strange goings on," small body bags and wreckage being carted away by military authorities. A second crash site was also claimed.

Others were highly skeptical. The former weather officer for the U.S. Air Force, Irving "Newt" Newton, who at the time identified the wreckage as that of a weather balloon, insisted that talk of flying saucers was foolish, "a bunch of horse pukey"!

However, in September 1994, after mounting pressure on the authorities, the Roswell incident was officially identified in a U.S. Air Force report as part of a secret American atomic spying program. The wreckage was probably a surveillance balloon launched as part of Project Mogul, a top-secret effort to detect Soviet nuclear weapon explosions. Charles Moore, a physicist in New Mexico, said that he believed the wreckage came from a six-hundred-foot craft as part of Project Mogul. The material in the debris was simply chemically treated balsa wood, and the strange markings were a variety of flower-patterned tape made by a toy company. In June 1997 a four-year, multimillion-dollar investigation by the Pentagon called *Roswell: Case Closed* confirmed the Project Mogul explanation, noting that the surveillance balloon trailed disk-shaped radar reflectors.

What then of the stories of alien bodies taken away by military authorities? It seems that for a few years after 1947 the air force conducted experiments that involved dropping dummies from high-altitude balloons to study the results of the impact. Witnesses' descriptions of the "aliens" closely match the characteristics of the dummies.

This continual changing of the story did nothing to give credibility to the official line, and indeed has fueled belief that aliens did crash at Roswell and the government covered it up. A recent investigation by the General Accounting Office found that important documents relating to the incident were missing, with the implication that they were destroyed more than forty years ago.

On the other hand, the flying-disk story itself could have been released as a coverup to conceal Project Mogul. Flying saucers were in the news, and a story could have been released either by mistake or on purpose.

In terms of the crash itself, one wonders how, having negotiated the difficulties of interstellar space and using such advanced technology that limits their appearance to a few instances here and there, such an alien spacecraft could be downed in a thunderstorm.

In addition, why do UFOs often appear so close to military bases? Those in favor of an alien origin will argue that aliens have revealed how worried they are by growing nuclear and military technology, and so keep a close eye on it. But why should they be worried? The technology they would have would be far more advanced than 1940s military capabilities, seen by the very fact that they traveled here. And why do they have to send craft down so low? Without the technology of extensive space travel we humans can see virtually all we need to know about things on a planet's surface from satellites in orbit.

The Roswell crash by itself is not convincing. But more was to follow.

Science-Fiction Video or Alien Autopsy?

Ray Santilli is a British producer of music videos. A few years ago Santilli was in the United States looking for early movies of 1950s rock stars when an elderly cameraman offered him something quite different. The man claimed that he had been called to Texas in July 1947 to film alien wreckage and an autopsy from the Roswell crash site. He had held on to twenty canisters of film and now wanted to sell them for a six-figure sum. Santilli bought the film.

This twenty-minute video shows grainy black-and-white footage of six-fingered, five-foot humanoids with huge heads, distended stomachs and no navels. The video was packaged and screened throughout the world. It looks quite real and convinced many people that aliens had indeed crashed at Roswell.

But on closer inspection there are number of problems with this video.

The bodies look oddly human. That is, they are not strange enough. Chris Stringer of the Natural History Museum in London wrote, "It's most improbable that aliens could have evolved to look so like humans. . . . It is my inclination to believe that the bodies are fakes, perhaps modified human bodies."[3]

The bodies could have been quite easily made artificially. Bob Keen, a special-effects expert on the movies *Alien* and *Judge Dredd,* likens the video to the effects being done for Hammer horror movies in the sixties. Special-effects experts who have studied the video point out that the posturing and weighting of the corpse was consistent with that of a body cast made in an upright position. Stan Winston, who did the special effects for *Jurassic Park,* although quoted by some as supporting the video, believes it to be a hoax.

There are difficulties with the cameraman's story. It is not clear why the cameraman was flown from Ohio to New Mexico, when experts would have been available much closer. It was against military procedure for the cameraman to develop the film himself, and indeed all procedures should have been shot in color rather than in black and white. During autopsies two cameras are usually used in a particular formation, and there should have been a stills photographer who would have been visible in the movie. Furthermore, the quality of camerawork is so bad that one can easily believe it was deliberately blurred in order to conceal the truth. Finally, the typed statement apparently from an American cameraman is full of phrases that suggest it was written not by an American but by an English person.

Finally, there has been no independent analysis of the film. Even Eastman Kodak, who had offered to date the film, has never received it.

Such problems severely undermine the authenticity of the Roswell film. Was it a hoax, or was it a real autopsy, not of aliens but the bizarre results of genetic engineering? We will not know until both the cameraman and the film come forward for independent examination.

Nevertheless, the story and belief in an alien crash at Roswell continue to gain momentum. After the video was released fragments of what were claimed to be the Roswell spaceship were sent to radio announcers from someone who said that he had picked them up as part of the retrieval team. Fragments were also handed in on March 24, 1996, at the UFO Museum at Roswell. Chemical analysis has so far revealed nothing unusual about these fragments.

What is clear is the "alien" crash pumps more than $5 million a year from tourism into Roswell as well as the massive spinoff of books and videos. The town is no doubt thankful for the crash, for without it it would be known only as having America's biggest mozzarella plant and being the birthplace of actress Demi Moore!

Government Coverups

The Roswell incident typifies the often widespread belief in government coverups. This is a major theme of *The X-Files* and crops up in relation to the Roswell incident in the film *Independence Day,* which follows the claim that the Roswell spaceship has been stored ever since its crash in a vault beneath Area 51, a secret Pentagon facility within the Nellis nuclear test range.

These conspiracy theories paint pictures of secret documents detained in underground vaults, government bureaus involved in covert UFO research and special departments that monitor military personnel in case of leaks of these top-secret documents.

Some theories take on bizarre aspects.

☐ Neil Armstrong and Buzz Aldrin encountered and photographed an alien craft on the moon in 1969, and the photographs were suppressed by the U.S. government. This has led to Armstrong's becoming a recluse, although why Aldrin seems relatively normal is not considered.

☐ The U.S. used alien technology to build bases on Mars and the Moon back in the 1950s. When John F. Kennedy decided to make this known to the American people, he was assassinated.

☐ The U.S. has set up top-secret Project Aquarius to gather data on alien life forms; Project Sigma to look at communications with aliens, which has led to a meeting between air force officials and aliens; and Project Snowbird, which involves test flying downed alien ships.

Apart from the lack of hard evidence to support these claims, the immediate question is: Why would the U.S. government do this? The theme of a deal between the government and aliens is a very popular one. The reason often cited is an agreement that allowed the U.S. military access to alien technology in return for aliens abducting humans. But where is such technology? Is the U.S. military using antigravity or phasers or warp drive? Why would aliens need to seek government approval to abduct humans? Surely they would be capable of doing it on their own.

Why would a government want to conceal the existence of aliens, and why would aliens want to keep their existence secret? Surely such a journey across the vast reaches of space would mean that they would want a state welcome. At this point people suggest the zoo hypothesis, that the aliens simply want to study us without being seen themselves. But this does not work. If you believe that there are aliens around because of the stories and sightings of UFOs, this then means that the aliens are not very good at keeping their existence hidden.

Much of this government-conspiracy feeling does come from a distrust of government power and secrecy. It is not beyond the track record of most governments in the world to use secrecy to develop technical advantages. Nor is it beyond them to use the "alien" story to cover up other military secrets. However, secrets have a habit of coming out. The evidence at present is not convincing that there is anything more than a government perhaps keeping an open mind on certain unexplained phenomena.

Alien Abductions

In 1966 John G. Fuller published a book entitled *The Interrupted Journey.* It was a sensation. It told the story of Betty and Barney Hill of New England, who believed that they had been abducted in New England and medically examined by aliens. The book encouraged a flood of abduction stories in America and then worldwide.

Up to this point, those who had seen aliens gave a wide range of descriptions. A survey claimed that

> about a fifth of the aliens were more or less humanlike; just over a third were small bipeds with huge heads; just under a third were not seen because of some clothing or helmet. Five percent were hairy bipeds. The remaining 11 or so percent were a miscellaneous bunch of complete weirdos.[4]

But abduction stories started to describe aliens who were all remarkably similar, and there were common characteristics to what happened. Abductions usually involved humiliating examinations and even sex with aliens.

This phenomenon seemed to affect a wide range of people. Aliens did not really seem to mind whom they abducted or where they abducted them. Debbie Jordan-Kauble was abducted by six aliens in 1983, and she claimed seeing bursts of light and power failures, markings left in her yard, having partial memories of what happened, and certain medical conditions such as unexplained rashes and irregular heartbeats.

All those who were abducted spoke of it being very real, of having a lasting impression on them and finding it difficult to talk about publicly. The use of hypnotic regression was often the key to unlocking these memories of abduction. The memories seemed to be repressed and could only be "released" with this technique.

Some exploited the interest. Whitley Strieber received a $1 million advance for the book *Communion,* which recounted his abduction in 1985. In 1993 he announced that he had actually never been abducted by aliens.

Budd Hopkins has interviewed many people under hypnosis and writes of it in his book *Intruders.* He announced that up to 3.7 million Americans have been abducted. Abduction books sell in the hundreds of thousands, and reports of different types of aliens and conspiracy theories abound. The common alien is the Grey, a figure about three feet tall with big black oval eyes who is part of the race that signed a treaty with Earth governments allowing them to abduct humans in exchange for alien technology.

Fire in the Sky?

Perhaps the most famous story of abduction concerns Travis Walton, told in the 1993 movie *Fire in the Sky.* Walton and six fellow workers were traveling home through a wooded area on November 5, 1975. They claim to have seen a UFO from their truck and stopped to check it out. Walton went to the clearing below the craft and a brilliant beam of light knocked him to the ground. For whatever reason the others drove off and called the police.

When they returned to the scene there was no trace of Walton or any physical evidence for a UFO. There was no sight of Walton for the next two days. Suggestions of murder were made, so the six were tested by a polygraph machine to see if their story was truthful. The polygraph indicated that five of the six seemed clearly not to be lying.

Five days later Walton called his sister from a gas station in a neighboring town, naked and upset. Although the details seemed difficult to ascertain, he believed that he had been abducted by aliens.

The story and indeed the movie seem very convincing. However, there a number of concerns that have been raised by those who believe in the existence of aliens and those who do not. First, Walton took a lie-detector test that came out negative, which seems to indicate that he was lying. The calm of his friends and Walton's own mother when confronted by news of the abduction seems to be a little disconcerting. Third, the truck driver with Walton admitted to having watched a movie on alien abduction two weeks before the incident happened. The workers were overscheduled and in financial crises, so a distraction would be beneficial, and the *National Enquirer* was offering a reward of $100,000 for proof of alien abduction. But there was no physical evidence at the abduction spot, and it might not be insignificant that Walton's father, who deserted him as a child, had been a spaceship fanatic.

Are Aliens Just in the Mind?

What really happened to Travis Walton may never be fully known. But what is beyond doubt is that many people truly believe that they have been abducted by aliens. This often happens while they're sleeping, and some claim to have been abducted up to three times per month. Some speak of the physical side effects that this causes and report that aliens put implants into their body during an experiment.

Are these things really happening? We must raise a question about the use of hypnosis to bring back repressed memories. In this form of regression therapy, hypnosis reveals what the patient believes to be true, not objective truth itself. Furthermore, such techniques are highly controversial in themselves. A person under hypnosis can sometimes be influenced by the person asking the questions.

Many within the UFO community itself are worried by the claims of four million abductions. Why would aliens need so many humans and deal with them one at a time? Using this set of figures tells us that being abducted by aliens is as likely as having a car accident! British insurance broker Simon Burgess is offering a policy that for a premium of £100 you receive £100,000 if you are abducted or £200,000 if you are sexually impregnated by aliens. He says that there is "significant demand" for such a policy. He has already sold three hundred policies and has had twelve thousand inquiries.

Few people, apart from insurance salesmen, appear to have researched this phenomenon, but recently there have been some scientific studies. The most famous was an extensive study done by John E. Mack, professor of psychiatry at Harvard.[5] He worked for three and a half years with over one hundred people who claimed they were abductees. Of these, some seventy-six fulfilled what he called the "abduction criteria," the conscious recall or recall with the help of hypnosis of being taken by aliens to a strange craft and having no apparent mental condition that would account for the story.

His conclusion was startling. He claimed that the abduction experiences were real, and he goes on to criticize a materialist worldview. He suggests that we participate in a universe or universes that are filled with intelligences from which we have cut ourselves off. As a result of this alienation the world has become subject to differences between rich and poor, violence, and ecological destruction. He describes the alien abduction phenomenon as having changed him profoundly and that it has the power to do so to others.

This sounds like religious language. Indeed, Christians would claim similar things. It is because we have cut ourselves off from the God of the universe that the world is in such a mess. For the Christian it is an encounter with the risen Jesus Christ that makes a profound change and has the power to do the same for others.

Mack is not alone in speaking in religious terms about alien abduction. Debbie Jordan-Kauble, mentioned earlier, says that since her abduction she has a new spirituality and belief in God. Another abductee, Elsie Oakensen, believes that since her abduction in 1978 she has been led to a healing ministry and a developing spirituality.

Are these people merely trying to reach beyond the dry, physical world of modern science and recognize the spiritual in life?

Whether it was for his religious language or his belief in aliens, Professor Mack, following the publication of his book, found himself the subject of a Harvard investigation into his work. This is almost unparalleled in recent academic circles.

Even without such a witch hunt there are a number of questions about his work. The problems with hypnosis have already been mentioned, and Mack has been criticized by colleagues for the way he gave his interviews and surveys. And unfortunately for Mack, one of his patients was a journalist on a debunking mission, which undermined Professor Mack's work to a degree.

But we are left with a number of options. Are all the abductees deluded or insane, is Professor Mack deluded or a fraud, or are the stories true?

Poached by a Spaceman or Paralyzed by Sleep?

There is another explanation. Sue Blackmore, a British psychologist, suggests that those who believe that they have been abducted by aliens are suffering from a phenomenon called sleep paralysis. She has gathered more than one hundred cases where she believes this to be the cause. The phenomenon occurs when a person is on the edge of sleep and lies semiconscious and aware but cannot move. Such paralysis normally occurs during dreaming as a natural safety belt to prevent us from acting out our dreams. In this state, dreams can seem like reality.

This theory fits with the victims' feelings of helplessness and the fact that many abductions happen during sleep or late at night. The people really do believe it happened to them, but in an objective sense there really were no aliens.

Others have suggested that the mind produces sightings of flying saucers. Michael Persinger of Laurentian University in Canada observed a correlation between flying-saucer reports and earthquakes. He speculates that the movement of the tectonic plates could release electromagnetic pulses that could then stimulate images in the mind. Such speculations are at a very early stage and need much more work. However, the fact that such speculations exist cautions us to be careful about putting too much weight on the UFO "evidence."

The evidence from cornfields and from abductions is not as strong as it might appear. There are still phenomena that seem to lie beyond our present ability to explain them. But to invoke aliens as a way of filling the gaps often raises more problems than it solves.

We have also touched on the idea of a link between religious themes and aliens. This has been developed by some as a way of finding more evidence of the presence of extraterrestrial life among us. It is to this that we turn next.

7

Was Jesus a Space Alien?

Has Earth been visited by aliens in its history? There have been many claims stating that we are descended from aliens or that monuments like Stonehenge and the statues of Easter Island were created by aliens. Perhaps the most provocative is the claim that religion is not the history of God's interaction with humans but the history and response to alien visitors.

Stanley Kubrick said in an interview with *Playboy* magazine in the 1960s,

> All the standard attributes assigned to God in our history could equally well be the characteristics of biological entities who billions of years ago were at a stage of development similar to man's own and developed into something as remote from man as man is remote from the primordial ooze from which he first emerged.[1]

Does the biblical record describe visits of aliens? In many books connected with questions of extraterrestrial intelligence such claims are made, for example, that the pillars of fire and cloud that led the Israelites to the Promised Land were in fact alien spacecraft or alien effects.

Even so careful and brilliant a popularizer as Paul Davies writes,

Indeed, it is easy to trace reports of flying craft and human-like occupants back into antiquity, where the reports merge with religion or superstition in a seamless manner. Consider, for example, the many Bible stories of angels coming from the sky, of humans ascending into heaven (the sky) or flying chariots. The most striking biblical account is perhaps that of Ezekiel who describes an encounter with four flying wheel-shaped craft "full of eyes" that "turned as they went" and out of which stepped the likeness of a man. The account may have been taken straight from a modern UFO report.[2]

Ezekiel's Close Encounter of the Chariot Kind

The claim that the Old Testament prophet Ezekiel saw an alien spacecraft was made popular by a NASA engineer, J. F. Blumrich, in his book *The Spaceships of Ezekiel,* published in 1974. Davies, like others, sees UFOs in the passage, citing evidence such as four flying wheel-shaped craft that were full of eyes, that turned as they went and out of which stepped the likeness of a man.

What does Ezekiel actually say? The following passage is from Ezekiel 1:

> I looked, and I saw a windstorm coming out of the north—an immense cloud with flashing lightning and surrounded by brilliant light. The center of the fire looked like glowing metal, and in the fire was what looked like four living creatures. In appearance their form was that of a man, but each of them had four faces and four wings. Their legs were straight; their feet were like those of a calf and gleamed like burnished bronze. Under their wings on their four sides they had the hands of a man. All four of them had faces and wings, and their wings touched one another. . . .
>
> Their faces looked like this: Each of the four had the face of a man, and on the right side each had the face of a lion, and on the left the face of an ox; each also had the face of an eagle. Such were

their faces. Their wings were spread out upward; each had two wings, one touching the wing of another creature on either side, and two wings covering its body. Each one went straight ahead. Wherever the spirit would go, they would go, without turning as they went. The appearance of the living creatures was like burning coals of fire or like torches. Fire moved back and forth among the creatures; it was bright, and lightning flashed out of it. The creatures sped back and forth like flashes of lightning.

As I looked at the living creatures, I saw a wheel on the ground beside each creature with its four faces. This was the appearance and structure of the wheels: They sparkled like chrysolite, and all four looked alike. Each appeared to be made like a wheel intersecting a wheel. As they moved, they would go in any one of the four directions the creatures faced; the wheels did not turn about as the creatures went. Their rims were high and awesome, and all four rims were full of eyes all around.

When the living creatures moved, the wheels beside them moved; and when the living creatures rose from the ground, the wheels also rose. . . .

Spread out above the heads of the living creatures was what looked like an expanse, sparkling like ice, and awesome. . . .

Then there came a voice from above the expanse over their heads as they stood with lowered wings. Above the expanse over their heads was what looked like a throne of sapphire, and high above on the throne was a figure like that of a man. I saw that from what appeared to be his waist up he looked like glowing metal, as if full of fire, and that from there down he looked like fire; and brilliant light surrounded him. . . .

This was the appearance of the likeness of the glory of the LORD.[3]

There is a lot more to this description than just a supposed spaceship. This picture is described not in terms of an historical narrative but as a "vision" when "the hand of the LORD was upon" the prophet. Similar experiences of seeing things of God are described in Ezekiel's own vision of a valley of dry bones (Ezekiel 37) and in other prophets such as Isaiah (Isaiah 6) and Daniel (Daniel 10). In none of these cases does anything like a spaceship appear.

Ezekiel clearly believes he sees a chariot-throne rather than a spaceship. Contrary to Davies, the vision does not describe "four flying wheel-shaped craft" but a rather complicated arrangement of living creatures and four wheels on which the chariot stood. Each wheel consisted of two wheels bisecting each other at right angles, thus allowing movement in any direction, although how they were attached to the chariot is an interesting engineering question. Further, these wheels were not flying but rose and descended with the living creatures. How the living creatures with four faces are a part of the UFO is always left out.

It is very difficult to translate or to understand what was originally meant by the phrase "full of eyes" (which has been thought to indicate some kind of portholes). The Scripture does not state that they "turned as they went," in fact quite the opposite.

Finally, the phrase "out of which stepped the likeness of a man" is somewhat misleading. The figure stays on the throne. And it was no mere humanoid that Ezekiel saw; Ezekiel qualifies the "figure like that of a man" with "the appearance of the likeness of the glory of the LORD."

By taking some elements out of context, reading into the particular verses things you want to see and frankly manipulating the words of the text to suggest something that it is not, it is possible to claim this was an alien spacecraft. No attention has been given to the rest of the passage and how that fits with the theory. Ezekiel was in exile in Babylon when he had this vision, and we know from the records that they left that the Babylonians were fascinated with objects in the sky and were good astronomers. So if this were such a stunning spacecraft, why did no one else see it?

In addition, no attention is paid to what the "alien" asks Ezekiel to do. He is to be a prophet to the people, bringing God's word to those Israelites who were in exile in Babylon. The message's content has little to do with cosmic philosophy but much to do with righteousness and sin, judgment and hope. Do these things really fit with an alien visitor?

This is the danger of inaccurate research. Texts are plucked from ancient documents without care to context or content, often interpreted in a particular way, and then these interpretations are repeated by author after author until they take on vast importance. Such manipulation of the facts is shown by the most famous author to suggest that aliens have long been visiting Earth, Erich von Däniken.

Chariots of the Gods and Other Miracles of Interpretation

In 1968 von Däniken published *Chariots of the Gods,* which became a worldwide bestseller, selling 3.5 million copies by 1970. He claimed that the Bible told the story of aliens who had visited our planet, or in other words, God was an astronaut.

Von Däniken suggested that aliens started the human race as a biological experiment, a form of genetic engineering. The story goes something like this: After a battle between two space peoples, the losers escaped to Earth. They eventually created human life "in their own image" from monkeys. After some time they left but still keep an eye on us. They came back in biblical times to lead Moses through the wilderness and built an interplanetary spaceport in Peru.

Von Däniken was a hotel keeper whose second book, *Return to the Stars,* was written while he was serving a prison sentence for embezzlement. It is estimated that 25 million people worldwide have read his claims. And in the 1990s there has been a renewed interest in his ideas.

Many authors have debunked von Däniken's claims and have shown the flimsy nature of the evidence he so confidently presents. For example, John Allan[4] points out a number of major problems.

1. Von Däniken's most famous claim is that alien visits are evi-

denced by huge markings on the Peruvian plain of Nazca. These miles and miles of strange markings, he argues, were runways for aircraft. But the plains are made out of soil, not a very solid runway, and aircraft would blow the markings away when they landed on them.

2. Von Däniken claims that maps belonging to a Turkish admiral, Piri Re'is, found at the start of the eighteenth century showed features that could only have been disclosed by aliens. These claimed features include accurate outlines of North and South America and mountain ranges in the Antarctic. On examination, however, these maps simply reflect what was known in the eighteenth century and are not as accurate as Von Däniken claims. For example, Japan is where Cuba ought to be.

3. He further states that a piece of cloth that was centuries old had to have been manufactured by aliens or with alien knowledge. But in fact it was made using the spinning and weaving techniques known at the time.

4. He maintains that the island of Elephantinos in the Nile was given this name because from the air it looks like an elephant, and only aliens could have seen this in ancient times. But the island was named as the site of an ivory market, the Greek *elephantinos* meaning ivory, not elephant.

5. Von Däniken claimed that a stone carving of a skeleton was so correct that it had to have come from alien x-ray techniques. Is it too simplistic to ask why the carver could not simply have looked at real skeletons?

6. Von Däniken's claims are often so vague that no one can check on his alleged facts, and he is at times factually wrong, as when he claimed that Enoch disappeared in a chariot of fire. If space allowed, we could go on.

Who Made the Chariots of the Gods?

Von Däniken invented and manipulated facts to suit his argument. His best pieces of evidence turn out to have straightforward scientific explanations. However, his argument suffers from another major flaw: If we were created by aliens, then who created them? If the reply is that their civilization arose naturally, then why did ours not arise naturally too?

Elsewhere I have referred to another claim that suffers from the same problem. Edward Harrison of the University of Massachusetts recently argued that we are here as the result of aliens, not in the sense of von Däniken or Crick, but in a larger cosmic sense. Recognizing that the universe is finely balanced for the emergence of life, Harrison believes that there are only three possible explanations. First, God designed it, but that answer, he argues, precludes further rational inquiry. Second, it appears to be designed by the very fact that we are here, but he finds this unsatisfactory. His third, and preferred, answer is that our universe was created by life of superior intelligence existing in another physical universe.

Thus this universe has brought forth life because it is an engineering project of superior beings. They created this universe out of a black hole. He calls this a "natural creation theory" and claims that it also explains why the universe is intelligible to us. It is created by minds similar to our own who designed it to be that way.

But where did these superior beings come from in the first place? Harrison criticizes belief in God for stopping any further rational inquiry, but then he falls into the same trap. What can we possibly know about these "superior" beings from another universe? If Harrison is drawn to the conclusion that our universe was designed, is it not simpler to see the "superior being" as God? Christians claim that this God, far from being in another universe, has revealed himself in this universe supremely in Jesus Christ.

To claim that theists say, "God created . . . ," and that this stops further inquiry is extremely naive. It was on the basis of belief in a Creator God that much early scientific revolution was begun. Far from stopping questions, belief in God can liberate inquiry.

Harrison's work, along with that of von Däniken, is significant because it is another example of just how contrived theories have to be to escape belief in God. In the last thirty years it has been fascinating to see many scientists being forced to ask theological questions as a result of discoveries about the universe,[5] questions such as

- Where do the laws of physics by which the universe originates and evolves come from?
- Why is the universe so finely tuned to allow the emergence of life?
- Why is the universe so intelligible to us?
- Why do we feel such awe at the universe that science discloses to us?
- What is the purpose of the universe?

Paul Davies and Sir Fred Hoyle have seen the answer to some of these questions as a sort of "divine" intelligence at work within the universe. They have no fundamental religious commitment (in fact, Hoyle was a leading atheist in the 1960s), yet they are saying, along with Harrison, that our knowledge of the universe means that we are not alone.

Of course all of these pointers about the universe are fully consistent with the Christian position that the universe has been created by a sovereign act of God. Indeed, this explanation, backed up by the claim that God has revealed himself in the space-time history of the universe supremely in Jesus, is much more powerful. However, what of the Christian claim that Jesus Christ is God's revelation? What if Jesus himself was an alien?

Beam Me Up, Father!

In von Däniken's defense, he never goes so far as to claim Jesus was an astronaut. In his 1974 book *Miracles of the Gods* he launches a sustained attack on the Roman Catholic Church and seems to suggest that Jesus is not important or advanced enough to be an alien visitor!

Nevertheless, von Däniken created a climate where it was a natural step to believe that many of the accounts of Jesus were simply the story of a supertechnological alien. The argument goes something like this:

- ☐ The virgin birth was really artificial insemination by an alien.
- ☐ Angels in "shining garments" were actually aliens in space suits.
- ☐ Jesus' saying "In my Father's house are many rooms" can only mean that there are many inhabited worlds in the universe.
- ☐ Miracles such as feeding five thousand people with a few loaves and fish were done by using devices like the food replicators in *Star Trek*.
- ☐ Walking on water was due to an antigravity beam.
- ☐ Prayer was really using a communicator with the spacecraft.
- ☐ The resurrection was achieved by the advanced medical science of the aliens.
- ☐ The ascension was simply "Beam me up, Scotty."

We need to ask a number of questions if these assumptions are to be discussed at all seriously.

Why Such a Life and Teaching?

There is a quality to Jesus as a human being that is attractive not only to the 2.5 billion Christians in the world today but also to many of those outside of the Christian faith, such as Gandhi. The Gospel accounts of his life picture a man in an obscure part of the Roman Empire who stood alongside the poor and the oppressed, healed the sick, and spoke the good news of God's forgiveness and love. As one Christian preacher put it, "Jesus sums up all that we should most of all like to be in our best moments." At the same time he spoke of God's judgment and the personal cost of being true to God's way.[6] It was a life of self-giving and a teaching that was firmly focused on God and his kingdom, rather than on space travel or other civilizations.

It is not difficult to argue that an alien could do this, but why *would* an alien do this? To show us a better way? The core of what Jesus is all about is love. Would an alien love us so much?

The Gospels are quite clear that Jesus was not primarily instituting a new social structure or even a code of personal ethics. He was offering a personal invitation. To see God one must look at Jesus. He did not just teach about truth, life, light and resurrection, but *was* those things. God's forgiveness is found through him, and the challenge is to follow him as a disciple.[7]

Some years ago C. S. Lewis used a famous argument to clarify the only possible things Jesus could have been: mad, bad or God. The depth and attractiveness of his life and moral teaching meant that it was impossible to condemn him as insane or a fraud, so that left just one alternative: Here was God himself walking the pages of history. Although many scholars would say that such an argument is somewhat simplistic, it still carries much force.

Why Such a Death?

Von Däniken himself asks whether we can believe that "the helpers from the spaceship came too late to save their top man from death. . . . If we imagine that space travelers would have left their important special messenger in the lurch, we are really underestimating extraterrestrial beings."[8]

But the point about Jesus' death is not that it was a mistake but that it was absolutely central to his mission. He spoke of his death often in terms of something that was both unavoidable and necessary.[9] What would the death of an alien achieve?

Roger Forster uses an illustration to oppose this line of thinking. Imagine two lovers walking on the bank of a river. Suddenly the man says to the woman, "To show how much I love you, I'm going to kill myself in the river," and he jumps into the river and drowns. What would the woman think? Probably that she was better off without him! Now imagine that they are walking along and the woman slips and falls into the river. The man jumps into the river and gets the woman to safety, but in the process he is carried off and drowns. What would the woman think then? She would know she was loved.

Christians understand the death of Jesus as a supreme demonstration of God's love, God in Jesus doing something for us. It was necessary because of the way we have turned against our Creator. Salvation is achieving reconciliation between creature and Creator.

Why Such a Resurrection?

One could argue that the resurrection was the type of thing that a doctor on a starship would accomplish day to day. In a book entertainingly titled *God Drives a Flying Saucer,* author R. L. Dione suggests, "It should be much easier for anyone to picture a UFO removing the slab from the tomb than to imagine an invisible, supernatural being shouldering it out of the way."[10]

The trouble is that this is not the whole story. The Gospels are quite clear that the resurrection of Jesus was not a simple resuscitation, but that there was now something different about the one they knew as Jesus. He no longer seemed restricted to the spatial and temporal constraints of the universe, for example, he could appear in rooms where the doors were locked. He promised his everlasting presence would be with the disciples, and after his ascension and the giving of the Spirit the disciples knew, along with countless other Christians through the centuries, Jesus' living presence in their lives.

It seems to me that to believe realistically that Jesus was an alien is rather like what happened to the Aristotelian universe. Aristotle believed in a universe where Earth was at the center and the planets moved in circles around it. But to explain the astronomical observations, more and more smaller circles, epicycles, were added. Eventually this theory had to be discarded for a totally different interpretation.

The Jesus-was-an-astronaut theory may sound initially attractive, but when looking at the accounts of Jesus, one has to give this alien more and more powers, and devise more and more motives and schemes, when it is truer to the observations to adopt a different explanation: he was God as a human being.

Where's the Evidence?

Underlying this discussion, and the discussion of UFOs in general, is a major question: What is the evidence for these claims? We have strongly criticized von Däniken in his use of evidence, but some say that Christianity itself is built on very flimsy evidence. The Heaven's Gate cult ridiculed Christianity, yet their leader claimed that he had been incarnated as Jesus two thousand years ago.

Recently an unusual full-page advertisement appeared in the London *Times.* It advertised a new book, *The Tomb of God,*[11] under the banner "A challenge to a 2,000 year-old religion demanding blind faith and centered on an event which did not happen." It makes the suggestion that the resurrection was false and that the body of Jesus is buried in a tomb on a mountain in France.

Richard Andrews, an ex-diver specializing in mine clearance, and Paul Schellenberg, a civil engineer, had indulged their interest in history and persuaded the book's publishers to give them an advance of half a million dollars. The claim they make is nothing new, having been around in different guises for some years. Indeed, it appeared in the book *The Holy Blood and the Holy Grail,* which also sold many copies.

On publication the book received reviews in the secular press such as "a load of codswallop," "bilge" and "little more than deluded amateurs with an advanced train-spotting mentality." The "evidence" the authors used consisted of eighteenth-century documents that, they argue, held a series of geometric clues to this French tomb, as did a seventeenth-century painting and a set of ancient buildings. They seemed to be unaware that there was a vast body of evidence from many previous centuries that was quite straightforward and much more reliable.

Nevertheless, books like *The Tomb of God* peddle the notion that Christianity does not rest on evidence, but just on blind faith. Nothing could be further from the truth. It is my conviction that the evidence for the Christian faith is much stronger than the evidence we reviewed in the previous chapter for the visits of aliens to Earth. Christianity is built on good historical foundations.

□ The Gospel writers, although selecting their material to bring out a particular perspective, are careful in describing historical detail such as political figures and geography, which can be checked by external sources.

□ The earliest copies of the Gospels that we have are from the early second century, which is unusually good in terms of historical research. By comparison, the oldest manuscripts of the historian Tacitus are dated eight hundred years after the original. (There have even been recent claims that a fragment of Matthew's Gospel can be dated at A.D. 50 to 60, although this still remains somewhat controversial.)

□ It does seem that all of the Gospels were written at the very latest in A.D. 90, although many scholars would suggest they were earlier, before A.D. 75. This means that there was a short period between the death of Jesus and the writing of the Gospels. Even during this time there are clear indications in the Gospels themselves of the power of oral tradition and the use of earlier written sources.

There is a vast body of scholarly research on the historicity of the New Testament documents, surveying both the internal evidence and sources external to the New Testament.[12] It is certainly the case that no other body of literature has undergone such a thorough going over, and apart from one or two ex-divers or civil engineers, its historicity is viewed as far more reliable than seventeenth-century geometric patterns. But this is not often acknowledged. Reports of seeing UFOs are often accepted without any rigorous investigation. Of course there are no autopsy videos of Jesus' body, but there are many kinds of conspiracy theories, none of which stand up to scrutiny.

However, Christianity does not rely solely on its historical basis. Common to Christians in different denominations in many different cultures of the world is the belief in a personal encounter with the risen Jesus. The continuity of this subjective experience with the historical basis is the key test of Christian claims to truth.

Donald MacKay, former professor in the research department of communication and neuroscience at the University of Keele, wrote,

> The basis of a Christian conviction of the truth of his faith is not that he has solved an intellectual riddle, but that he has come to know a living Person—the Person of Jesus Christ. It is his new relationship with God that makes the doctrine ring true, not the other way round. The reason that so many of us lack this conviction is, I think, not that the evidence is not available, but that we look for the wrong kind of evidence in the wrong way. We look for facts and arguments, instead of being prepared to be met by a Person. At least, so it was with me.[13]

This experience can be likened in some way to alien abductions in the sense that here is an experience that goes beyond our scientific understanding and shows us graphically that we are not alone. But there the similarity stops. This is not an experience to be recalled under hypnosis; the individual is aware of it in a rational way at the time it happens. It occurs both day and night, in mundane, everyday situations and in extreme circumstances. It is not tied to sleep paralysis. Such experience has stood up over the centuries to scrutiny, ridicule and persecution. Most of all, it challenges a person to change.

This combination of evidence and experience is at the very least strong enough to merit serious examination. I believe that it has far stronger evidence than the evidence for aliens outlined in the previous chapter.

Yet it is still not proof. But that should not worry us. The UFO community asks us to make a judgment on the evidence available. Christianity does the same, but goes further. We do not have to wait to be one of the few who are "abducted," but we can experience the reality by opening our life to God now. This does not involve hypnosis, but a simple trust, a change of mind and a commitment to follow his will.

8

Some Alien Problems for God?

Although the evidence for extraterrestrial life and intelligence is scant, there is still a question to which we still must turn: Would the discovery of life elsewhere in the universe so contradict the central beliefs of Christianity that it would bring it crashing back to the grave?

The Christian church has been very silent on the idea of the possible existence of extraterrestrial life and intelligence. This might be due to ignorance about the scientific arguments involved or, perhaps, a fear that examining such a question would severely undermine Christianity. However, for the vast majority of the Christian church the existence of extraterrestrial intelligence is not a big deal.

Ernest Barnes, Bishop of Birmingham, in his Gifford Lectures in Edinburgh (1927-1929), suggested that there are likely to be many inhabited worlds. His argument depended on three ideas that we discussed in a slightly different guise earlier in the book. First, God created the universe for the emergence of consciousness, therefore consciousness would not be confined to just one world. Second, it is likely that there are many planets like our own in such a vast universe. Third, since the origin of life can be explained by physical processes, there is nothing special in the emergence of life.

We have seen so far that these assumptions are perhaps not as firm as Bishop Barnes would like. Nevertheless, there is quite a clear implication in his first assumption that there is nothing wrong in looking to God to extend his creativity to other beings throughout the universe.

Oxford cosmologist E. A. Milne went further: "Is it irreverent to suggest that an infinite God could scarcely find the opportunities to enjoy himself, to exercise His godhead, if a single planet were the seat of His activities?"[1]

Yet Paul Davies seems to think that extraterrestrial life and especially extraterrestrial intelligence constitute a major headache for God. It would overturn the "traditional religious view." This runs throughout his recent book, *Are We Alone?* and influential articles in *Reader's Digest* and *Time* magazines.[2] In this chapter I have extracted Davies's major criticisms to see whether they pose a threat. He believes that extraterrestrial intelligence raises problems for the Christian in terms of

1. humanity's special relationship with God
2. the origin of life
3. highly advanced alien religions
4. the significance of the life and death of Jesus

Humanity's Special Relationship with God

Johannes Kepler once wrote, "How can all things be for man's sake? How can we be masters of God's handiwork?" Davies sees the existence of extraterrestrial intelligence as somehow undermining what he sees as Christianity's claim that men and women have a special and exclusive relationship with the Creator.

In what sense does Christianity see human beings as "special"? This is not a question just for religion. The question of the difference between human beings and other life is an important one for the biological sciences and has ethical consequences as well.

For example, is it acceptable to perform experiments on animals? This has become a very controversial topic in recent years. While most people would condemn the unnecessary suffering of animals merely to perfect certain perfumes, it is not so clear when it comes to experiments designed to alleviate human suffering. In 1921 Frederick Banting and Charles Best experimented on dogs in a process that led to the discovery of insulin and thereby relief for millions of diabetics. Was this justified?

There are some biologists who argue, although they sometimes do not follow it through to animal rights, that the idea of neo-Darwinian evolution means that the special worth of human beings cannot be maintained. After all, they would say, accountants are part of a long chain leading to ashes of dead stars. We are made of the same stuff and produced by the same process as animals, so what is so special about us?

On this view we are "nothing but iron enough for one medium-sized nail, sugar enough to fill seven cups of tea, lime enough to whitewash one chicken coop, phosphorus enough to tip 2,200 matches, magnesium enough for one dose of salts, fat enough for seven bars of soap, potash to explode one toy crane, and sulfur to rid one dog of fleas."

We may be all this, but is that all we are? Donald MacKay pointed out many years ago that this "nothing buttery" was reductionism in the extreme and just did not work. For as things in this universe get together, they become more than just their parts. You cannot reduce a human being to nothing but the same chemicals as everything else. Most people recognize this. However, it is a little more difficult to actually categorize what differences there are.

How are we human beings different from other forms of life on this planet? Amoebas are very different from humans, but is there something that distinguishes us from the higher mammals?

Our genetic codes differ by less than 2 percent from those of a chimp. Sensitivity to pain and a capacity for intelligent behavior are elements often mentioned, but animal studies show that these are differences of degree rather than kind. Other differences between animal and human nature have been suggested:[3]

- the ability to learn, plan and conceptualize
- the use of developed tools, language and counting
- an artistic sense
- the ability to integrate a wide range of different areas of knowledge
- the ability to make intuitive acts of judgment (which is often seen in science)
- moral sense
- the ability to recognize one's self in a mirror (which is also shared by chimps and blue whales)
- the capacity for language and abstract thought, which brings with it power to reflect on pain and death
- the ability to understand abstract mathematics and then use such mathematics to ask fundamental questions, such as about the origin of the universe

There have been attempts to explain some of these things within an evolutionary framework. For example, E. O. Wilson[4] has argued that genes determine our social behavior. So altruism, instead of being a "higher God-given quality," is preprogrammed into us by our genes so that the individual sacrifice helps the tribe (or gene pool) to survive. The unselfish person helps the group.

Even if this idea were generally accepted, which it is not, we need to be careful to read Wilson's scheme as "nothing buttery." Even if the process by which some of these things come about can be understood, they still mark human beings as different from the rest of life on Earth.

There is a further, current aspect of this debate. When Deep Blue, a computer, beat Gary Kasparov at chess, was this a sign that computers demonstrate intelligence comparable to the human level? And if so, then what really is the difference between Luke Skywalker and his droid C3PO?

Faced with these questions and reacting against the view that we are nothing but part of nature and the evolutionary chain, some people stress our total distinctiveness. This view sees the origin and many aspects of human beings as beyond the ability of science to describe. But what does the Bible actually say about this? Does it take the line that human beings are totally distinct from everything and are therefore special in the sight of God?

First of all we need to recognize that the creation accounts in the book of Genesis stress the continuity between human beings and the rest of creation. For example, in the week of creation, both animals and humans are created on the same day (Genesis 1:24-31), and in the more specific second chapter, "the LORD God formed the man from the dust of the ground" (Genesis 2:7). That man was created from dust stresses the relationship that human beings have with the rest of God's creation (cf. Genesis 2:19).

In one sense we are part of the same creation and creative process as the rest of life on this planet. Therefore we should not be worried that we share much in terms of our biochemistry and some of our behavior patterns with animals. Indeed, it is because we are an integral part of the natural world that we can use science on ourselves.

However, the Bible does not say that we are "nothing but" dust. If the first part of Genesis 2:7 points to the origin of human beings in nature, the second part points to something that marks us out as different: "the LORD God . . . breathed into his nostrils the breath of life, and the man became a living being." The phrase "living being" is also used of animals, but the picture here is of God directly giving his breath, a picture of intimate relationship, quite different from the rest of creation.

This is put more forcefully in Genesis 1:26-28:

> Then God said, "Let us make man in our image, in our likeness, and let them rule over the fish of the sea and the birds of the air, over the livestock, over all the earth, and over all the creatures that move along the ground."

> So God created man in his own image,
> in the image of God he created him;
> male and female he created them.
> God blessed them and said to them, "Be fruitful and increase in number; fill the earth and subdue it. Rule over the fish of the sea and the birds of the air and over every living creature that moves on the ground."

Human beings alone are given a privileged and responsible position. We alone are made in the image and likeness of God, with considerable responsibility to rule wisely over the rest of God's creation. We are of nature and have dominion over nature. What makes us special is this relationship with God.

The Image of God. Theologians have struggled over the centuries to understand what the Bible means when it says that human beings are made in the "image of God." There are several suggestions about what *image* means:

☐ a physical resemblance to God
☐ morality and rationality
☐ a receiving of dominion over Earth, which involves stewardship and creativity
☐ self-awareness and reflective self-consciousness

The view of most contemporary Old Testament scholars is that the image of God should not be thought of as a "part" of us. It is not about something we have or something we do, but it is about relationship. Old Testament scholar Claus Westermann writes, "Human beings are created in such a way that their very existence is intended to be their relationship to God."[5] The image of God means that we are sufficiently like God that we can have an intimate relationship with him.

This is emphasized often in the Genesis account. God "walks" in the Garden with Adam and Eve, and he speaks in a different way to them than he does to other created beings. He speaks personally, and they understand and respond.

This is how the Bible understands the special nature of human beings, not that we are physically different from the rest of creation, though in many ways we are, but that God has given us an intimate relationship with himself. It is this relationship that marks us out from the animal kingdom. Mark Twain said, "Man is the only animal that blushes, or needs to." The breaking of this relationship, which ruins the good, is where our moral sense comes in.

Unique but Not Exclusive. The special nature of human beings does not come just from the Christian understanding of creation. Most of all it comes from Jesus. God's special love toward human beings is shown most strongly by what Christians call the doctrines of incarnation and redemption. Incarnation is God's becoming a human being in Jesus and living among us in the space-time history of the universe. Redemption is God's dying a human death to restore that intimate relationship that we had destroyed by turning away from him. It is because of these acts of God that the special nature of human beings is not undermined by the existence of other life.

The Christian faith is already used to dealing with plenty of nonhuman life in the universe. From amoebas to elephants, Earth itself is teaming with nonhuman life. This has not caused Christianity any major problems. Biblical faith sees this natural world as part of God's rich creation, where he exhibits diversity, extravagance and beauty. Bacteria on Mars would simply be part of this great creation tapestry.

Westermann comments,

> The simple fact that the first page of the Bible speaks about heaven and earth, the sun, the moon and stars, about plants and trees, about birds, fish and animals, is a certain sign that the God whom we acknowledge in the Creed as the Father of Jesus Christ is concerned with all these creatures, and not merely with humans. A God who is understood only as the god of humankind is no longer the God of the Bible.[6]

God's care is never exclusive. Humanity may be unique in its relationship to God, but uniqueness does not imply exclusiveness. I have a unique and special relationship with my wife, but that does not mean that we do not relate to any other person at all. There are different degrees of friendships and family relationships with many people. Even relationships at the same level can be unique and special. We have a unique relationship with our son that is very special indeed. However, that is not to say that we do not have an equally special but different relationship with our daughter.

God is a God of relationship. In Colin Gunton's words, "God is love in advance." His very being as Trinity, experiencing and giving love within three persons, demonstrates that supremely. His incarnation in Jesus of Nazareth is a visible sign of the love that breaks through walls of racism, nationalism and sexism, a lesson that the people of Israel had to learn over and over again. The nation of Israel, chosen by God for particular purposes, enjoyed a special relationship with him. That did not mean, however, that God's purposes were restricted just to that nation. God's love was for all, for Jew and Gentile alike.

We can go further. The value and care of the environment is central to the Bible. It is a mistake to ignore this and believe that human beings alone have value, a mistake that has partly led to our environmental crisis. We should not make a similar mistake in terms of extraterrestrial life and intelligence.

If such life exists, then it has value to God, the creator of all things. From bacteria on Mars to aliens in flying saucers, we have a responsibility to respect and find out about them, for we could learn more about God's extravagance in creation. This is a biblical prime directive.

Extraterrestrial intelligence does not pose a problem to the Christian belief that men and women are special in the eyes of God. It may even increase the sense of awe at how great this God is who loves his creatures so much.

Colin Russell, professor of the history of science and technology at the Open University, makes a fascinating point in relation to this.

He points out the popularity in the seventeenth century of speculating about other worlds. This he sees as a significant indicator of the ascendancy of biblical values over Aristotelian. In the Aristotelian universe, position and status were closely associated. Earth was at the center of all things, separated from the rest of universe by the orbit of the Moon. We were special because we were placed at the center. In contrast, the Bible does not associate status with place. The dignity and worth of human beings comes from the gift of relationship with God.

Human beings are not the center of the universe. In fact, it is the human belief that we are the center of all things that the Bible calls sin. It leads to the arrogant treatment of the rest of the created order and the breaking of human relationships. Some writers give the impression that our destiny is to control the universe, but that is not the biblical view. God is the center of all things, and we are creatures given status by his love. As a recent article rightly states, "We are not the central focus of all that is. All life reflects God's glory, not ours."[7]

The Origin of Life

Davies's second problem for Christianity in relation to extraterrestrial intelligence is the origin of life. He believes that Christians hold to the idea that life comes from a unique, divine miracle. If the Earth is not unique, then this would undermine the case for a miraculous origin, and even for God. What Davies fails to distinguish is that a divine origin for intelligence or consciousness can happen within the physical process and can be unique without being exclusive.

There are those within the Christian church who say that we are special because of the mechanism by which we were physically created. This is the same view that thinks the early chapters of Genesis give a scientific description of the creation of human beings. It was a miraculous and distinct creative act that brought about the first man and woman, so evolution is false, or at least evolution between species.

This is a view I respect but I do not share, for primarily biblical reasons that I detail elsewhere.[8] Nevertheless, it is difficult to see Davies's problem even in this context. There is no reason at all why God cannot miraculously create whatever he wants to elsewhere in the universe. Indeed, it might be easier for those sharing this view to cope with the arrival of other intelligent life than it would be for those who believe in evolution, who would be hurriedly reexamining their probability equations.

Other Christians who are equally committed to the authority of the Bible are quite prepared to accept that God's creative action might have been through a process rather than a single act. From the time of Darwin some Christians have argued that evolution was God's mechanism for producing life, a mechanism sustained and guided by God.

Such people did not reject evolution (but are still cautious due to its incompleteness) because they understood from the Bible that the special place of humanity was not predicated on the terms of our creation. God had given us the capacity for intimate relationship and had demonstrated it so graphically in the death of Jesus in order to restore that relationship. This gave us freedom to follow the biblical mandate to practice science and to test what science was saying without being threatened.[9]

What Makes Adam, Adam? If evolution is accepted, what would the Christian picture be? Although we accept that human beings are to a large degree more advanced than anything else on the planet, the key to being human is to be in relationship with God.

The evolutionary process may have produced the physical characteristics of humankind, but at a certain point God reached out to us in relationship. He does this throughout the rest of the Bible by calling Abraham, Moses, the prophets and many, many more. The New Testament gives us the picture of God as the one who takes the initiative and reaches out to re-form a broken relationship.

God formed a relationship with the first human beings, Adam and Eve. They are the first human beings because of this intimate, responsible relationship that God holds out. According to anthropologists, creatures with physical features similar to modern human beings existed at least 100,000 years ago. Civilization, however, is much younger, probably beginning between 8000 and 6000 B.C. The Genesis account places Adam and Eve in this period in the part of the Middle East where the earliest civilization developed.

Taking this view, the problem that Davies raises does not seem to be at all serious. God is the source and sustainer of the physical process, which he can be either on this planet or on another. We can still affirm that life and intelligence have a divine origin. Incidentally, as we saw in chapter three, Davies's own preference for a "law" that favors the growth of complexity leading to life begs the question of where such a law itself comes from. Is this, along with the other laws of physics that make life possible, a pointer toward a divine origin of a universe capable of bringing forth life? Instead of being a problem for God, the discovery of extraterrestrial life as Davies envisions it becomes a pointer toward God.

Certainly scientists who are Christians have nothing to fear in looking for signs of life elsewhere in the universe. It is part of our biblical mandate to explore. This is our "continuing mission."

Highly Advanced Alien Religions

Davies's third problem for traditional religion is the expectation that if alien communities exist, many of them will be far more advanced than we are, and therefore their religion would be better than ours. After millions of years of development they would have become far more enlightened in philosophy, science, wisdom, knowledge and religion. Davies writes,

> God's progress and purposes will be far more advanced on some other planets than they are on Earth. . . . It might be the case that aliens had discarded theology and religious practice long ago as primitive superstition and would rapidly convince us to do the same.

> Alternatively, if they retained a spiritual aspect to their existence, we would have to concede that it was likely to have developed to a degree far ahead of our own. If they practiced anything remotely like a religion, we should surely soon wish to abandon our own and be converted to theirs.[10]

Davies adopts this view because he has little concept of revelation. He sees religion as intellectual progress, so that another society more intellectually advanced would be bound to have a higher religion. But Christianity's central claim is not a spiritual evolution that will take us closer and closer to God as our knowledge increases. It realistically acknowledges how finite our minds are in the face of the infinite and is based on the fact that God reveals truth about himself in a way that we can understand. As time goes on our knowledge of God grows, but the foundation of this is God's particular revelation.

Our fundamental need is not a superreligion but a reconciliation we cannot achieve for ourselves. God achieves it through a particular action in a particular place and moment in time.

Biblical Christianity sees revelation and salvation inextricably linked in the life, death and resurrection of Jesus of Nazareth. Christians should therefore not be afraid of the religion of alien beings, just as they should not be afraid of human beings from the different faith communities in the world. The belief that God has revealed himself in a supreme way frees us to look for him.

Christians would learn new things about God from an encounter with aliens, but they would also be in a position to share the good news that God has revealed himself in becoming a human being and has offered us salvation.

If highly advanced alien religions are not a problem for biblical Christianity, they do raise Davies's final point: How can the particular revelation and act of reconciliation of Jesus of Nazareth be linked to life elsewhere in the universe? Would extraterrestrial intelligence show that the great claims that Christians make for Jesus are simply wrong?

The Significance of the Life and Death of Jesus

The Bible sees the death and resurrection of Jesus as central to God's purposes for the whole universe. Davies thinks this is a problem in terms of what the discovery of extraterrestrial life would mean. As another writer puts it, "How can the Christian gospel, concerned with the salvation of men in this world, have any universal significance when we know that there may well be intelligent life on other planets?"[11] How can Jesus Christ have more than an earthly significance? Can aliens know God, and if so, how can they be saved?

The Bible does see Jesus Christ in a cosmic sense.[12] He is the Lord over creation who makes loaves and fish feed the five thousand, who stills the storm, who heals those with physical needs. The people cry, "What kind of man is this? Even the winds and the waves obey him!" (Matthew 8:27).

The confident answer of the New Testament is that he is not only Lord over creation but Lord of creation. He is the one through whom the whole universe was brought into being. This is part of an early and widespread Christian belief and is represented in three key passages in the New Testament (John 1:1-18; Colossians 1:15-20; Hebrews 1:1-14). The Jesus encountered by the first-century fishermen of the Sea of Galilee is the one through whom and for whom the universe was created. He is the eternal Son, the Word of God.

The New Testament is clear that at the same time he is Lord over creation, he is also part of creation. Alongside these cosmic views of Christ, his humanity is also stressed. He shares our human nature and dies on the cross (Philippians 2:6-11).

It is this tension that underlies any discussion of the relationship of Jesus Christ to extraterrestrial life. Although not many Christian thinkers have addressed this question, some have and have then gone two different ways.

1. Once for All? Some believe that God's incarnation in Jesus was the one particular event in the whole universe where God showed his nature most fully and provided the mechanism for reconciliation. On this understanding we would be commissioned to preach the good news of this event not only to the nations but to the galaxies. Cosmologist E. A. Milne observes,

> God's most notable intervention in the actual historical process, according to the Christian outlook, was the Incarnation. Was this a unique event, or has it been re-enacted on each of a countless number of planets? The Christian would recoil in horror from such a conclusion. We cannot imagine the Son of God suffering vicariously on each of a myriad of planets. The Christian would avoid this conclusion by the definite supposition that our planet is in fact unique. What then of the possible denizens of other planets, if the Incarnation occurred only on our own? We are in deep waters here in a sea of great mysteries.[13]

Milne eventually solves his great mystery by suggesting that we send the good news by radio waves.

While this would no doubt be attractive to some televangelists, it still is major problem. As we made clear in chapter four, these missionary messages would take a very long time to reach their audience, never mind the possibility of the missionaries themselves reaching the aliens. This factor should not be underestimated. Would God allow some intelligent life in the universe to be physically barred from hearing this supreme revelation of himself?

Nevertheless, this is not a new problem for the Christian faith. Theologians have long wrestled with the question of what about those who, because of where or when in world history they were born, do not have the opportunity to hear the gospel of Jesus. The question of aliens is simply an extension of this.

Without giving a full answer, the Bible stresses that some revelation of the character of God can be seen in the creation itself and that a person can be saved through Christ without having heard of Christ. For example, Abraham, who lived a long time before Jesus was even born, is called

God's friend (James 2:23-24). He was justified by faith in the grace of God, looking forward to God's particular act, unlike Christians today who are put right by faith looking back on that act. Aliens beyond the reaches of communication could follow a similar pattern.

But alongside this acknowledgment, the particularity of the Christian gospel has always been a major motivation for telling others the good news. Sometimes this has been done with arrogance and oppression quite out of character with its message. It has also been done with humility and courage. Perhaps faster-than-light travel will be possible someday and the Christian church will have fresh opportunities to tell the good news.

2. God in Little Green Flesh? Some Christians have taken a different opinion on the Incarnation as it relates to the universe. They have suggested that God would take the form of other intelligent life in the universe and become incarnate on different planets in different ways. Milne's view was severely criticized by theologian E. L. Mascall in his Bampton Lectures in 1956. He argued that if salvation is what God was all about, then he would have made sure his creatures knew about it. Mascall stressed that salvation has to be achieved through incarnation. That Jesus became man means it is doubtful that his saving work would be for different types of beings.

Mascall prefers a different idea:

> If there are, in some other part or parts of the Universe, rational corporeal beings who have sinned and are in need of redemption, for those beings and for their salvation the Son of God has united (or one day will unite) to his divine Person their nature, as he has united it to ours.[14]

Such a view has many advocates. Christian rock singer Larry Norman, in his song *UFO,* wrote,

> And if there's life on other planets,
> Then I'm sure that he must know.
> And he's been there once already,

And has died to save their souls.[15]

From a completely different style of music and part of the church, hymn writer Sydney Carter says much the same.

Who can tell what other cradle,
High above the Milky Way,
Still may rock the King of Heaven
On another Christmas Day?

Who can count how many crosses
Still to come or long ago
Crucify the King of Heaven?
Holy is the name I know.

Who can tell what other body
He will hallow for his own?
I will praise the Son of Mary,
Brother of my blood and bone.

Every star and every planet,
Every creature high and low,
Come and praise the King of Heaven,
By whatever name you know.[16]

These songs are simply expressing a truth at the heart of biblical Christianity: God loves and, because of his love, acts. That there is no limit to that truth allows it to be transferred to the speculation of extraterrestrial intelligence.

Close Examinations of the Theological Kind

However, some Christians, like E. A. Milne, look on such a prospect of multiple incarnations and indeed multiple sacrificial deaths with suspicion. There is a theological disagreement between Christians on this matter, which can be highlighted in three questions.

1. What is the relationship between Jesus the man and Jesus the eternal Word of God? Theologian Norman Pittenger argued that many find unnecessary difficulty with the view of multiple incarnations because of "Jesu-centrism," the belief that the human life of Jesus in Palestine is thought to give complete knowledge of God. However, the significance of Jesus is as the incarnation of the Eternal Word of God, the second person of the Trinity who becomes flesh in man. This Jesus Christ is central and decisive for our human relationship with God—showing what God is like, what human beings are meant to be, the way of reconciliation and restoration of our true nature. But he is not the whole of what God is about. "For Christian faith, Jesus defines but does not confine God in his relationship to the created world."[17]

Pittenger leaves open the question of whether God takes on the flesh of other intelligent life, but he argues that the basic truth is that we would expect God to show the same interest, care and judgment on other worlds. We must believe that what God reveals in Christ is in continuity with what he is doing elsewhere. Jesus Christ is our clue to all God does anywhere and everywhere. Frank Weston, Bishop of Zanzibar, echoed this as early as 1920:

> If other planets support rational life . . . I am quite certain that Christianity is revealed to them in some way corresponding with its revelation to us. Our Christianity is the self-unveiling of eternal Love in terms and forms intelligible to us . . . their Christianity will be the self-unveiling of eternal Love in terms and form intelligible to them. . . . It is only those who erect a false barrier between the universal activity of the Word and his incarnate life as a man who will boggle at the possibility of his self-revelation in a created form on another planet.[18]

The importance of this view is that it emphasizes that incarnation and salvation are basic to who God is. It reminds us that self-giving love is at the heart of God.

It has its dangers as well. To drive a wedge too far between the "cosmic Christ" and the "human Jesus" does begin to open the door to the view that Jesus was just a good man used by God. This is not true to the biblical data. So some Christians who note this danger do not want to go down this road.

2. Do aliens sin? Such a view as expressed by Pittenger and Mascall links revelation and salvation together. Does salvation have to be linked with incarnation, or can God save in other ways? Our knowledge of this is limited to one case—us! This link between salvation and incarnation is due to the fact that we have fallen from God's way and so simultaneously need revelation and salvation. C. S. Lewis made this point when he suggested that if an alien race does exist, then we are left with the question of whether it has fallen in the same way as human beings.

At an even more basic level this raises the question of the doctrine of the Fall. This is the biblical understanding that we are not as we should be, that we have fallen from God's purposes and ideals.

Some Christians would stress that the Fall occurred at a particular point of history on Earth and that it had consequences for the whole of the created order. Would other intelligent life be affected by the fall of Adam and Eve? The difficulty is that there is just not enough evidence within the Bible to make confident claims one way or another. That the created order is not as it should be and that human beings are not as they should be is quite clear. However, how the Fall affects other intelligent life cannot be fully spelled out.

3. Where will Jesus return? A further objection to the view of God's becoming incarnate on a number of planets concerns the relationship of the cross of Christ to God's saving work for the whole of creation. Is it the New Testament view that the whole creation, not merely the human race, is saved by Christ's work on the cross? Present theological thinking puts a major emphasis on God's relationship to creation. This is partly a reaction against a position that sees humans being saved by being "plucked out" of this creation.

This is not the biblical view of God's purposes. There are clear indications that God's work is not just for human beings but for the reconciliation of all things and for the creation of a new heaven and earth.[19] This will happen when Jesus returns in glory.

Some Christians take this to mean that there is only one Incarnation and that we are alone in the universe because these things are so tightly tied to human beings. However, we must recognize that many biblical passages have little interest in the rest of the universe apart from us. The often-quoted (but difficult to interpret) Romans 8:18-25 speaks of the whole creation groaning in travail, awaiting redemption from futility, but the redemption of the children of God is at the center of the picture.

Those who believe in many incarnations would probably say that the return of Jesus would be such a momentous event that it would be seen not just on Earth[20] but throughout the universe. This must be the case. If it were not, the Lord of heaven and earth would be a rather small deity, and his purposes would not reach the whole of creation.

Speculation or Necessity?

At this point we must remind ourselves that we are speculating and certainly unable to answer all the questions fully. As one child wrote on a school science exam, "To most people solutions mean finding the answers. But to chemists solutions are things that are still all mixed up." Some of our solutions may still seem all mixed up. Christians will come to different conclusions as they weigh the biblical evidence and will just have to honestly admit ignorance to some questions.

We must always recognize that the nature of God is such that he will do whatever is required. John Polkinghorne, president of Queens College, Cambridge, said, "If little green men on Mars need saving, then God will take little green flesh. . . . He will do what is necessary."[21] How God reveals himself elsewhere in the universe is an interesting speculation, but the clear emphasis of the Bible is how we respond to what he has revealed to us here.

The question of the relationship of Jesus Christ to extraterrestrial life is not an easy one, but Christian theologians have grappled with it. There are just too many unknown factors to be specific. That is not avoiding the question, but rather being honest with respect to our present knowledge. Perhaps to find out the full picture we will have to wait until we have contact with aliens.

Stargate

In one of the memorable scenes from *Independence Day,* a group of UFO enthusiasts cluster on a tall skyscraper with banners of welcome held up to the large spaceship above their heads. But these confident people are the first ones destroyed by the energy beam from the spaceship.

What should the reaction of Christians be if we were ever visited by extraterrestrial life? Davies seems to think that the church should immediately institute a closing-down sale. But as we have seen, his arguments are weak and do not in any way constitute a destructive beam of reasoning. I do not agree that "it is hard to see how the world's great religions could continue in anything like their present form should an alien message be received."

I believe that the discovery of extraterrestrial intelligence would be exciting for the Christian, for it would open up even more of the glory and stunning creativity of the God revealed to us in Jesus. That is not to say that it would not pose difficult questions, but Christianity is used to dealing with difficult questions from the world of science. Many of these difficult questions have ultimately enriched our faith rather than destroyed it.

Perhaps Christians should be at the forefront of some of the research into extraterrestrial intelligence. Roman Catholic theologian Stanley Jaki[22] suggests that it is only the theist who can look forward with confidence to such an encounter, trusting that both sides will have a common Creator and a sense of brotherhood.

9

The Truth Is in There Somewhere

Columnist Bernard Levin wrote in 1995, "If you just think for a moment about those vast numbers of other worlds, you should be rocking with laughter if anyone suggests that the Universe is peopled only by us."[1]

If you are rocking with laughter at this point, I will humbly suggest that we have not fully understood the uncertainty and complexity of the arguments concerning the existence of extraterrestrial intelligence. We have reviewed the scientific evidence, the claims of aliens visiting our planet and the theological questions. The theological "problems" suggested by Paul Davies are not serious enough to influence a conclusion. The evidence for alien spaceships and abductions, although it does raise some intriguing questions, is not conclusive.

The scientific evidence, however, is at present pessimistic about the existence of extraterrestrial intelligence, although it is slightly more open to the possibility of extraterrestrial life. We may have discovered planets outside our solar system, but many more factors are needed to produce life on those planets. Our understanding of the biological origin of life and the development of intelligence seems to point to the probability that we are alone in the galaxy, if not the universe. If there is a "law" of increasing complexity that results in intelligence, then our view might be altered, but this is very much speculative at this stage.

The lack of any meaningful radio signal and Fermi's "if they existed, they would be here" argument reinforce this pessimism of discovering intelligent life on another planet. We must always bear in mind that there may be intelligence so far from us that no contact would ever be possible, in which case we would never know that it was there.

We have pointed out at each stage the uncertainties in all of the individual arguments, but taken together they form a strong case against the existence or knowledge of alien intelligence. Of course new discoveries can change the picture, the confirmation of independent evolution of life on Mars being a big step. However, we must keep emphasizing the distinction between extraterrestrial intelligence and life. Life could be plentiful in the universe in the form of bacteria, for example, but it would not necessarily imply intelligence. Indeed, unless we stepped on it, we would not be aware of it! And certainly bacteria are not able to build radio telescopes. In a real sense we would still be alone in the universe.

My own view is that at present there is no strong evidence for extraterrestrial intelligence, and scientific arguments are strongly against it. That is not to say that there may be life of sorts in the universe, but we cannot even be confident of that. As a scientist and a Christian I want to encourage the search for extraterrestrial life and intelligence. If it is there, it would help our understanding of some physical processes that we do not as yet fully understand. It would also demonstrate more of the extraordinary creativity of God.

Why Do We Want to Believe?

If present evidence is so against extraterrestrial life, why then does

Bernard Levin want to rock with laughter? His language of "other worlds" suggests that he has already made up his mind. Is it a cool scientific assessment of the arguments involved, or is it the expression of something much deeper, the human spirit crying out that we cannot be alone in the universe? The latter is strongly felt by many. We have seen it in science fiction and in the quest of many to show that we are not alone. Why do we so want to believe?

There are good historical and sociological grounds for why there has been an amazing growth of belief in the existence of extraterrestrial intelligence in the latter part of the twentieth century. World War II ignited the belief in UFOs. People's direct experience of seeing rocket technology suggested for many, including the scientists who worked on the projects, that space travel was just around the corner, and in many ways it was. The Apollo program put a man on the moon and returned him safely to Earth. Space travel has been taken for granted ever since. From the Moon to Mars and from Mars to the Milky Way seemed to be the inevitable result of this progress. After nearly three more decades we still have not yet set foot on Mars, but the idea is still there.

Alongside this came the claim that we were on the verge of understanding the origin of life. The discovery of DNA and the seeming ease with which Miller and Urey produced amino acids in 1953 gave the impression that life could arise spontaneously throughout the universe.

In this last decade of the twentieth century we have caught "millennium fever." The immanence of millennium has a psychological impact. At the turn of the first millennium prophets of doom interpreted the signs to proclaim that the end was at hand. The current decline of traditional Christianity in the West has contributed to the search for alternatives. The heightened sense of the spiritual has been expressed more and more in the belief in aliens as the millennium approaches.

It is also clear that the belief in aliens is linked to the rise of New Age thinking. Many of the magazines that report UFO sightings and abduction cases couple with them paranormal and occult events. The collection of ideas under the title "New Age" provides fertile ground for various beliefs, the most common being the rejection of the materialistic and mechanistic worldview so dominant since the time of Newton. This opens up an interest in things that lie beyond conventional science's ability to give answers. The expression of a universal religious sense finds outlets in paranormal experiences and contact with aliens. The New Age protests that there is more to the universe than just an improbable evolution of a species totally alone in a mechanistic, impersonal universe.

I believe that there are a number of fundamental components underlying the question "Are we alone?" that are more important. They pick up on concerns that have been fundamental throughout the history of human beings.

Lost in Space

The first concern is a feeling that can be described only as cosmic loneliness. In one of the more famous phrases of modern science, Jacques Monod wrote, "Man at last knows that he is alone in the unfeeling immensity of the universe." Monod is following the logic of the improbability of the origin of life discussed in chapter three.

If we are alone in the universe, then the vastness of the universe is not just awe-inspiring, but almost chilling. David Hughes writes,

> The confirmation of the existence of extraterrestrial life is billed as the greatest possible scientific discovery of all time. Today, however, we are still experiencing the pangs of cosmic loneliness. Never mind not coming to visit, no extraterrestrial being has even left a calling card or shouted at us from a distance. What is even more enigmatic is the realization that it is just as amazingly incredible to insist that Earth is the only repository for sentient life forms in the

> Universe as it is to envisage the hoards of other inhabited planets orbiting billions of distant stars.[2]

To accept that we are alone in the universe makes us face the cosmic fact that we are one planet revolving around an insignificant star in an insignificant galaxy in a universe of 100 billion stars in each of 100 billion galaxies. Greta Garbo once famously proclaimed, "I want to be alone," but many cannot cope with that. We are frightened to believe that we are the best and the only life that the universe can produce.

Stephen Spielberg's classic movie *ET* was popular for a number of reasons, not least because children loved the fact that here was an extraterrestrial who was fun, friendly and the same height as they were. Not only were we not alone, but we had friends out there who were able to ride on bicycles with us! Yet it was not enough to be just a friend. ET came from the skies, healed people and was eventually raised from the dead. He bore striking similarities to the Christian belief in Jesus as friend and risen Savior.

Is there something in us that is looking for a cosmic friend? Science-fiction author Ray Bradbury wrote after the NASA claim of life on Mars,

> This latest fragment of data . . . is only worth our hyperventilation if we allow it to lead us to the larger metaphor: Mankind sliding across the blind retina of the Cosmos, hoping to be seen, hoping to be counted, hoping to be worth the counting.[3]

Why Are We Here?

Another human concern is trying to discover the purpose of the universe. Are we so unique that the purpose of the universe is in some way linked to us? Or could aliens share some secret knowledge with us? Physicist Stephen Weinberg joined the cosmic pessimism of Monod when he commented that the more the universe seems comprehensible, the more it also seems pointless.

Paul Davies argues that this pessimism is a result of the belief that the processes of nature are essentially random, which is compounded by the fact that science tells us that the end of the universe is doomed to futility, either in a fiery death or in the collapse of the universe into the opposite of a Big Bang—a Big Crunch. Philosopher Bertrand Russell follows this line in the book *Why I Am Not a Christian.*

> The world which science presents for our belief is even more purposeless, more void of meaning. . . . That man is the product of causes which had no prevision of the end they were achieving; that his origin, his growth, his hopes and fears, his loves and his beliefs, are but the outcome of accidental collocations of atoms, that no fire, no heroism, no intensity of thought and feeling, can preserve an individual life beyond the grave; that all the labors of the ages, all the devotion, all the inspiration, all the noonday brightness of human genius, are destined to extinction in the vast depth of the solar system, and the whole temple of man's achievement must inevitably be buried beneath the debris of a universe in ruins—all these things, if not quite beyond dispute, are yet so nearly certain that no philosophy that rejects them can hope to stand.[4]

Does human life count for nothing? Is there really no purpose to our place in this vast universe? Russell accepts that human life ultimately counts for nothing, a chilling thought. Some think that we do not ask the question of life's purpose, but in reality most of us do. The belief in extraterrestrial life is a way of getting beyond that. Paul Davies sums it up: "For those who hope for a deeper purpose beneath physical existence, the presence of extraterrestrial life forms would provide a spectacular boost, implying that we live in a universe that is in some sense getting better and better rather than worse and worse."[5]

Here pessimism is replaced with optimism, although it is hard to see why. Other life in the universe may give a sense of purpose in terms of Davies's "law" that produces life, but it is still very impersonal.

Nevertheless, it may give some a sense of purpose. John Allan commented in 1975, "There is growing public hunger for something to believe in—something which combines the certainties of science with a religious optimism about the future that science on its own cannot justify."[6] Belief in extraterrestrial life can do that.

Who Are We?

A third concern for human beings is that we want to find out about ourselves. We do that fundamentally in relationship. Science fiction has used this device on many occasions. *Star Trek* reflected American culture of the 1960s in its exploration of themes such as racism through encounters with "aliens." We want to find out about aliens because we want to find out about ourselves. The recent TV hit comedy *Third Rock from the Sun* uses such a device. A group of aliens take human bodies and enable us to laugh at ourselves as we see ourselves from their perspective.

Aristotle said, "All men by nature desire knowledge," but we often want knowledge in relation to our own place in the universe. How did we get here? Are we unique? Would alien life continue the process of dethroning the centrality of humanity, begun with Copernicus taking Earth from the center of the universe and carried forward with the claim of evolutionism that we are just a random product?

Live Long and Prosper?

We all seem to have some sense of "cosmic fear." H. G. Wells's novel *The War of the Worlds* was written in response to the outrage he felt at the colonialist eradication of the people of Tasmania. His aim was to show what it was like to be a victim of a war of extermination.

However, Orson Wells's 1938 radio version of the book had quite a different effect on the American public. It produced widespread fear and panic in many Americans, who were in the grip of prewar paranoia. Science fiction works on such fear and paranoia. The fear of being attacked by aliens is the dominant theme in movies from *Invasion of the Body Snatchers* to *Alien* to *Independence Day*. Indeed, *Independence Day* pictures the nightmare scenario: vastly superior aliens come with the evil purpose of eradicating the entire human race from Earth. These movies feed on the fear and vulnerability induced by a world outside our own and outside our control. We feel our vulnerability.

There Must Be a Better World Somewhere

Finally, we want to be saved. One correspondent to *Sightings* magazine commented on what an encounter with aliens would mean:

> Their positively vast medical technology could result in new medical techniques, ridding us of cancer or even AIDS. [They could] possibly resolve our planet's famine and drought problems, and allow us to open our eyes and minds and discover the rest of the universe.

As far back as 1949, Sir Fred Hoyle pointed out that this might be one motivation for believing in extraterrestrial intelligence, "the expectation that we are going to be saved from ourselves by some miraculous interstellar intervention."[7] The hope for many is that something outside of ourselves will come and save us from the reality of life. We look beyond our present knowledge for hope. This was represented in an extreme form by the Heaven's Gate suicide. One woman said in her farewell video, "I've been here on this planet for thirty-one years, and there's nothing here for me."

Once again, this is often reinforced by science fiction. Rick Berman, executive producer of the *Star Trek* spinoff series *The Next Generation, Deep Space Nine* and *Voyager,* commenting on the series' phenomenal popularity, said, "*Star Trek* offers a future that is better than the present." This was fundamental to the belief of its original creator, Gene Roddenberry. The characters of *The Next Generation* were to have grown beyond insecurity and aggression, although why they then needed a full-time ship's counselor is difficult to understand.

The Truth Is Here

These fundamental questions are not confined just to UFOs and radio telescopes. They are questions that have been addressed throughout the centuries by philosophy and theology. Paul Davies rightly sees that the interest in extraterrestrial intelligence "stems in part . . . from the need to find a wider context for our lives than this earthly existence provides. In an era when conventional religion is in sharp decline, the belief in super-advanced aliens out there somewhere in the universe can provide some measure of comfort and inspiration for people whose lives may otherwise appear to be boring and futile."[8]

This comfort and inspiration is not just an individual belief. Those who believe in aliens attach themselves to a subculture of language, ideas and media. This gives community, security and a feeling of superiority. People think they can gain secret knowledge in a world where knowledge is growing at such a quick rate.

This UFO community can sometimes be a substitute religion. We see this in extreme cases in new religious cults that place aliens at center stage, such as the Church of Scientology.

What is fascinating is that questions of loneliness, purpose, identity, fear and salvation arise. They are an expression of a universal sense that we are not alone in the universe. Where does that come from? As we have seen, the evidence is not strong in favor of there being other intelligent life in the universe. Some, following Freud, would simply say that these things are just our psychological inability to cope with the truth that we are alone. We are the "crippled child crying out in the night" in a universe where we are alone.

There is an alternative view. It is that "we need a wider context for our lives" because we were created that way. If God created us to be in an intimate relationship with him, but we have broken that relationship, it is no wonder that we are obsessed with questions of loneliness, purpose, identity, fear and salvation. Augustine put it this way: "God made us for himself, and our hearts are restless until they find rest in him." We know that we are not meant to be alone. French existentialist and atheist Jean-Paul Sartre put it clearly: "That God does not exist I cannot deny, but that my whole being cries out for God I cannot forget."

We are crying out, but we are crying out for the wrong thing. It is rather like a child crying desperately to play with her toy when in fact what she really needs is a good nap.

The needs for companionship, purpose, self-understanding, reassurance and help to make things better are all real. They are an expression of men and women who are made in the image of the Creator of the universe yet feel alienated from God and their true selves.

The Christian faith gives answers to these questions. We are not alone. The God who made the universe wants to be in relationship with us. There is a purpose to our existence. We are created as an act of extravagant love by God. We can understand ourselves by reference to and experience of that love. That same love is a source of security, and out of that love God came in Jesus to offer salvation here and now.

Is this wish fulfillment? No. As we have seen, there is good evidence for this, focused supremely in the life, death and resurrection of Jesus of Nazareth. There is no proof, but the evidence points to an invitation to make personal contact. This involves making a commitment on the basis of trust. Such a commitment is demanding. On God's side it means forgiveness, companionship and his own power given to the life of the Christian. On our side it means opening up our lives to be changed and following God's way of self-giving love.

The belief in aliens is just scratching the surface. It is a means of expressing deep religious convictions without the commitment and truth to actually encounter reality. Extraterrestrial life may exist, a prospect that would be exciting for Christians. But such life will never deliver answers to loneliness, purpose, identity, fear or salvation.

Are we alone in the universe? If God's existence is not obvious, it is easy to feel that we are alone and to look for security in something else. Two thousand years ago the Jewish people under the political oppression of the Roman Empire looked for a messiah, a deliverer to come with force to usher in God's kingdom. They followed a stream of would-be messiahs, all who failed.

Then a child was born to an everyday couple in a town called Bethlehem, a child who was given the name Immanuel, which means "God with us." Matthew tells the story of this child. He came not with force but with love, healing the sick and standing with the oppressed. He taught the kingdom of God not in terms of war but in terms of peace. He did not achieve political power but was put to death on a cross.

Matthew records the words Jesus cried out from the cross, "Eloi, Eloi, lama sabachthani?" and then gives the translation, "My God, my God, why have you forsaken me?" On the cross Jesus experiences what it means to be totally alone. God in Christ experiences our alienation in order to reconcile us back to himself.

The result of this achievement is shown in the words of the risen Jesus, which form the conclusion to Matthew's Gospel.

> Then the eleven disciples went to Galilee, to the mountain where Jesus had told them to go. When they saw him, they worshiped him; but some doubted. Then Jesus came to them and said, "All authority in heaven and on earth has been given to me. Therefore go and make disciples of all nations, baptizing them in the name of the Father and of the Son and of the Holy Spirit, and teaching them to obey everything I have commanded you. And surely I am with you always, to the very end of the age."[9]

We are not alone in the universe.

Notes

Chapter 1: Is the Truth Out There?

[1]Paul Davies, *Are We Alone?* (London: Penguin, 1995), p. xi.

[2]Colin Russell, *Cross-Currents: Interactions Between Science and Faith* (Leicester, U.K.: Inter-Varsity Press, 1985), p. 52.

[3]N. West, "The Cult of the X-Files," *Sightings* 1, no. 2 (1996): 38.

[4]Timothy Good, *Beyond Top Secret* (London: Sidgwick and Jackson, 1996).

[5]Nick Pope, *Open Skies, Closed Minds* (London: Simon & Schuster, 1996).

Chapter 2: Close Encounters of the Scientific Kind

[1]Frank Drake and Dava Sobel, *Is Anyone Out There?* (London: Simon & Schuster, 1994), p. 233.

[2]Quoted in *Focus,* April 1995, p. 41.

[3]See, for example, I. S. Shklovskii and Carl Sagan, *Intelligent Life in the Universe* (San Francisco: Holden Day, 1966); F. D. Drake, *Intelligent Life in Space* (New York: Macmillan, 1960); P. Morrison, "Conclusion: Entropy, Life and Communication," in *Interstellar Communication: Scientific Perspectives,* ed C. Ponnaperuma and A. G. W. Cameron (Boston: Houghton Miffin, 1974); A. G. W. Cameron, ed., *The Search for Extraterrestrial Life* (New York: Benjamin, 1963).

[4]See, for example, T. Dobzhansky, *Genetic Diversity and Human Equality* (New York: Basic Books, 1973), p. 99; T. Dobzhansky et al., *Evolution* (San Francisco: Freeman, 1977); François Jacob, "Evolution and Tinkering," *Science* 196 (1977): 1161; E. Mayr, "Evolution," *Scientific American* 239 (September 1978): 46.

[5]D. S. McKay et al., "Search for Past Life on Mars: Possible Relic Biogenic Activity in Martian Meteorite ALH 84001," *Science* 273 (1996): 924.

[6]Quoted in *The Daily Telegraph,* January 27, 1996.

[7]The Kelvin temperature scale (K) is equivalent to the Celsius scale but has a different zero point. It begins at absolute zero, which is -273° Celsius. Thus the freezing point of water is 273 Kelvin, and its boiling point is 373 Kelvin. Of course, at high temperatures the difference is not too important.

[8]Michael D. Lemonick, *Time,* February 5, 1996, p. 47.

Chapter 3: It's Life, Jim, but Not As We Know It

[1]Karl R. Popper in *Studies in the Philosophy of Biology,* ed. F. J. Ayala and T. Dobzhansky (London: Macmillan, 1974), p. 259.

[2]Fred Hoyle and Chandra Wickramasinghe, *Evolution from Space* (London: J. M. Dent & Sons, 1981).

[3]Francis Crick, *Life Itself* (London: MacDonald, 1981).

[4]"Which Is the Chicken and Which Is the Egg," editorial, *Nature* 372 (1994): 29.

[5]J. N. Hawthorne, *Windows on Science and Faith* (Leicester, U.K.: Inter-Varsity Press, 1986), p. 68.

[6]See, for example, Phillip Johnson, *Darwin on Trial*, rev. ed. (Downers Grove, Ill.: InterVarsity Press, 1994).

[7]For example, Richard Dawkins, *The Blind Watchmaker* (Harmondsworth, U.K.: Penguin, 1988); *The Selfish Gene* (Oxford: Oxford University Press, 1989); *River out of Eden* (London: Weidenfeld and Nicholson, 1995).

[8]E. Mayr, *One Long Argument: Charles Darwin and the Genesis of Evolutionary Thought* (London: Allen Lane, 1991).

[9]Dawkins, *Blind Watchmaker,* p. 5.

[10]John D. Barrow and F. J. Tipler, *The Anthropic Cosmological Principle* (Oxford: Oxford University Press, 1986), p. 129.

[11]C. O. Lovejoy in *Life in the Universe,* ed. J. Billingham (Cambridge, Mass.: MIT Press, 1981), p. 326.

[12]M. Ruse, *Darwinism Defended* (Reading, Mass.: Addison-Wesley, 1982).

[13]D. C. Dennett, *Consciousness Explained* (Harmondsworth, U.K.: Penguin, 1993).

[14]D. C. Dennett, *Kinds of Mind: Toward an Understanding of Consciousness* (London: Weidenfeld & Nicolson, 1996).

[15]D. Chalmers, *The Conscious Mind* (Oxford: Oxford University Press, 1996).

[16]Stuart Sutherland in *Macmillan Dictionary of Psychology* (Basingstoke, U.K.: Macmillan, 1995), p. 95.

[17]Paul Davies, *Are We Alone?* (London: Penguin, 1995), p. 84.

[18]Stuart Kauffman, *The Origins of Order* (Oxford: Oxford University Press, 1993).

Chapter 4: Calling Occupants of Interplanetary Craft

[1]Karen Carpenter and Richard Carpenter, "Calling Occupants of Interplanetary Craft," Klaatoons, ATV Music, © 1977. Used with permission of Music Sales.

[2]Carl Sagan, *Contact* (London: Arrow Books, 1988).

[3]D. W. Hughes, review of *Extraterrestrial Intelligence,* by J. Heidmann, *The Observatory* 116 (1996): 183.

[4]G. Lake, "V/V_{Max} and the Detection of Evolving Extraterrestrial Intelligence," *Quarterly Journal of the Royal Astronomical Society* 33 (1992): 357.

[5]P. Wesson, "Cosmology, Extraterrestrial Intelligence and a Resolution of the Fermi-Hart Paradox," *Quarterly Journal of the Royal Astronomical Society* 31 (1990): 161.

[6]Frank Drake and Dava Sobel, *Is Anyone Out There?* (London: Simon & Schuster, 1994), p. 233.

Chapter 5: To Boldly Go Where No Alien Has Gone Before

[1]J. von Neumann, *Theory of Self-Reproducing Automata,* ed. and comp. A. W. Burks (Urbana: University of Illinois Press, 1966).

[2]John D. Barrow and F. J. Tipler, *The Anthropic Cosmological Principle* (Oxford: Oxford University Press, 1986), p. 576.

[3]W. I. Newman and Carl Sagan, "The Solipsist Apprach to Extraterrestrial Intelligence," *Quarterly Journal of the Royal Astronomical Society* 24 (1983): 113.
[4]J. A. Ball, "The Zoo Hypothesis," *Icarus* 19 (1973): 347.

Chapter 6: Crop Circles, UFOs, Abductions and All That

[1]Jim Schnabel, *Round in Circles* (London: Hamish Hamilton, 1994).
[2]Charles Berlitz and William L. Moore, *The Roswell Incident* (London: Granada, 1982).
[3]Quoted in the London *Times,* July 29, 1995.
[4]P. Harpur, *Daimonic Reality: A Field Guide to the Otherworld* (London: Viking Arkana, 1995), p. 23.
[5]John E. Mack, *Abduction: Human Encounters with Aliens* (New York: Simon & Schuster, 1995).

Chapter 7: Was Jesus a Space Alien?

[1]J. Angel, ed., *The Making of Kubrick's "2001"* (New York: New American Library, 1970), pp. 331-32.
[2]Paul Davies, *Are We Alone?* (London: Penguin, 1995), p. 87.
[3]Ezekiel 1:4-28.
[4]John Allan, *The Gospel According to Science Fiction* (London: Falcon, Church Pastoral Aid Society, 1975).
[5]David Wilkinson, *God, the Big Bang and Stephen Hawking* (Crowborough, U.K.: Monarch, 1996).
[6]See, for example, Luke 4:14-21; 5:12-26; 7:37-42; 9:18-27; 15:1-32; 16:19-31.
[7]See, for example, John 3:1-21; 6:35-59; 8:12; 14:5-14.
[8]Erich von Däniken, *Miracles of the Gods* (Ealing, U.K.: Corgi Books, 1977), p. 100.
[9]See, for example, Mark 8:31—9:1, 2-13; 10:32-34.
[10]R. L. Dione, *God Drives a Flying Saucer* (New York: Bantam, 1973), p. 123.
[11]Richard Andrews and Paul Schellenberg, *The Tomb of God* (Boston: Little, Brown, 1996).
[12]See, for example, F. F. Bruce, *The New Testament Documents: Are They Reliable?* (Downers Grove, Ill.: InterVarsity Press, 1982); F. F. Bruce, *Jesus and Christian Origins Outside the New Testament* (London: Hodder, 1984); Craig Blomberg, *The Historical Reliability of the Gospels* (Downers Grove, Ill.: InterVarsity Press, 1987); J. D. G. Dunn, *The Evidence for Jesus* (London: SCM, 1985); R. Forster and V. P. Marston, *Christianity, Evidence and Truth* (Crowborough, U.K.: Monarch, 1995).
[13]Donald MacKay, *The Open Mind* (Leicester, U.K.: Inter-Varsity Press, 1988), p. 17.

Chapter 8: Some Alien Problems for God

[1]Quoted in Paul Davies, *Are We Alone?* (London: Penguin, 1995), p. 30.
[2]Paul Davies, "The Harmony of the Spheres," *Time,* February 5, 1996, p. 52.

[3]For example, W. H. Thorpe, *Animal Nature and Human Nature* (London: Methuen, 1974).
[4]E. O. Wilson, *Sociobiology: The New Synthesis* (Cambridge, Mass.: Harvard University Press, 1975).
[5]Claus Westermann, *Genesis 1-11* (London: SPCK, 1986), p. 158.
[6]Ibid., p. 176.
[7]W. Clarke, *Baptist Times,* August 15, 1996.
[8]David Wilkinson, *God, the Big Bang and Stephen Hawking* (Crowborough, U.K.: Monarch, 1996).
[9]David Wilkinson and R. Frost, *Thinking Clearly About God and Science* (Crowborough, U.K.: Monarch, 1996).
[10]Davies, *Are We Alone?* pp. 33-37.
[11]W. N. Pittenger, *The Word Incarnate* (London: Nisbet, 1959), p. 248.
[12]Colin Gunton, *Christ and Creation* (Carlisle, U.K.: Paternoster, 1992).
[13]E. A. Milne, *Modern Cosmology and the Christian Idea of God* (London: Oxford University Press, 1952), p. 153.
[14]E. L. Mascall, *Christian Theology and Modern Science* (London: Logmans, 1956), p. 36.
[15]Larry Norman, "UFO." Reproduced by permission of Cyril Shane Music Ltd.
[16]Sydney Carter. Copyright 1961 Stainer and Bell Ltd. Used by permission.
[17]Pittenger, *Word Incarnate,* p. 249.
[18]F. Weston, *The Revelation of Eternal Love* (London: Mowbray, 1920), pp. 128-29.
[19]See, for example, Ephesians 1:10; Revelation 21.
[20]See, for example, Matthew 24:26-27; 1 Thessalonians 4:13-18.
[21]Quoted in *The Observer,* August 11, 1996.
[22]Stanley Jaki, *Cosmos and Creator* (Edinburgh: Scottish Academic Press, 1980).

Chapter 9: The Truth Is in There Somewhere

[1]Bernard Levin, "Outer Space, Out of Mind," *Times* (London), August 22, 1995.
[2]D. W. Hughes, review of *Extraterrestrial Intelligence,* by J. Heidmann, *The Observatory* 116 (1996): 183.
[3]Ray Bradbury, "Our Martian Destiny," *The Wall Street Journal,* August 21, 1996.
[4]Bertrand Russell, *Why I Am Not a Christian* (New York: George Allen and Unwin, 1957), p. 107.
[5]Paul Davies, *Are We Alone?* (London: Penguin, 1995), p. 52.
[6]John Allan, *The Gospel According to Science Fiction* (London: Falcon, Church Pastoral Aid Society, 1975), p. 39.
[7]F. Hoyle, "On the Cosmological Problem," *Monthly Notices of the Royal Astronomical Society* 109 (1949): 365.
[8]Davies, *Are We Alone?* p. 89.
[9]Matthew 28:16-20.

For Further Reading

Allan, John. *The Gospel According to Science Fiction.* London: Falcon, Church Pastoral Aid Society, 1975.

Barrow, John D., and F. J. Tipler. *The Anthropic Cosmological Principle.* Oxford: Oxford University Press, 1986.

Billingham, John, ed. *Life in the Universe.* Cambridge, Mass.: MIT Press, 1981.

Crick, Francis. *Life Itself.* London: MacDonald, 1981.

Crowe, Michael. *The Extraterrestrial Life Debate.* Cambridge: Cambridge University Press, 1986.

Davies, Paul. *Are We Alone?* London: Penguin, 1995.

Drake, Frank, and Dava Sobel. *Is Anyone Out There?* London: Simon & Schuster, 1994.

Forster, R., and V. P. Marston. *Christianity, Evidence and Truth.* Crowborough, U.K.: Monarch, 1995.

Gunton, Colin. *Christ and Creation.* Carlisle, U.K.: Paternoster, 1992.

Heidmann, Jean. *Extraterrestrial Intelligence.* Cambridge: Cambridge University Press, 1995.

Houghton, J. T. *The Search for God: Can Science Help?* Oxford: Lion, 1995.

Hoyle, Fred, and Chandra Wickramasinghe. *Evolution from Space.* London: J. M. Dent & Sons, 1981.

Jaki, Stanley. *Cosmos and Creator.* Edinburgh: Scottish Academic Press, 1980.

Kauffman, Stuart. *The Origins of Order.* Oxford: Oxford University Press, 1993.

Klass, Philip. *UFOs Explained.* New York: Random House, 1974.

Lucas, E. *Genesis Today.* London: Scripture Union, 1989.

Mack, John E. *Abduction: Human Encounters with Aliens.* New York: Simon & Schuster, 1995.

MacKay, Donald. *The Open Mind.* Leicester, U.K.: Inter-Varsity Press, 1988.

Menzel, Donald H., and E. H. Taves. *The UFO Enigma.* New York: Doubleday, 1977.

Papagiannis, Michael, ed. *The Search for Extraterrestrial Life.* Dordrecht, Netherlands: Reidel, 1985.

Polkinghorne, John. *Quarks, Chaos and Christianity.* London: Triangle, 1994.

Pope, Nick. *Open Skies, Closed Minds.* London: Simon & Schuster, 1996.

Russell, Colin A. *Cross-Currents: Interactions Between Science and Faith.* Leicester, U.K.: Inter-Varsity Press, 1985.

———. *The Earth, Humanity and God.* London: UCL Press, 1994.

Schnabel, Jim. *Round in Circles.* London: Hamish Hamilton, 1994.
Wilkinson, David. *God, the Big Bang and Stephen Hawking.* Crowborough, U.K.: Monarch, 1996.
Wilkinson, David, and R. Frost. *Thinking Clearly About God and Science.* Crowborough, U.K.: Monarch, 1996.
Zuckerman, Ben, and Michael H. Hart, eds. *Extraterrestrials: Where Are They?* Cambridge: Cambridge University Press, 1995.

Index